Municipal Solid, Agricultural, and Mining Waste in Sub-Saharan Africa

The book seeks to cover the application of the principles of the circular economy within the context of the processing and waste management industry in sub-Saharan Africa. It includes topics related to the circular economy and the sustainable use of resources in the mining, process industry, and waste management sectors, and examines how some resources from these sectors, currently treated as "waste", can be converted into secondary resources that can help drive economic development.

Features:

- Describes fundamental environmental and process engineering principles to understand waste valorization technologies.
- Covers sustainable solid waste management and recovery of resources from solid waste from an African perspective.
- Provides guidelines for specific approaches to long-term sustainable waste management.
- Includes alternative waste treatment technologies in order of priority as per the waste hierarchy.
- Introduces pertinent industry-specific case studies, worked examples, and problems.

This book is aimed at researchers and graduate students in environmental and chemical engineering and waste treatment.

Africa Circular Economy Series

CRC Press (Taylor and Francis Group) introduces **Africa Circular Economy Series** (with a Science and Technology outlook but preferably more interdisciplinary). Under this initiative, we invite scholars, academicians, researchers, and professionals to contribute to this series. We are interested in discussing monographs, references, textbooks, shortform books and handbooks to add to our book programs, at undergraduate, postgraduate and doctoral level. All the books published under the series will be part of our Global Publishing program.

Circular and Transformative Economy: Advances towards Sustainable Socio-economic Transformation
Edited by Luxon Nhamo, Sylvester Mpandeli, Stanley Liphadzi and Tafadzwanashe Mabhaudhi

Municipal Solid, Agricultural, and Mining Waste in Sub-Saharan Africa
Sustainable and Circular Valorization Opportunities
Freeman Ntuli, Tebogo Mashifana, Thandiwe Sithole, and Thabo Falayi

Municipal Solid, Agricultural, and Mining Waste in Sub-Saharan Africa

Sustainable and Circular Valorization Opportunities

Freeman Ntuli, Tebogo Mashifana, Thandiwe Sithole, and Thabo Falayi

CRC Press
Taylor & Francis Group
Boca Raton London New York

CRC Press is an imprint of the
Taylor & Francis Group, an **informa** business

Designed cover image: chayanuphol\Shutterstock

First edition published 2025
by CRC Press
2385 NW Executive Center Drive, Suite 320, Boca Raton FL 33431

and by CRC Press
4 Park Square, Milton Park, Abingdon, Oxon, OX14 4RN

CRC Press is an imprint of Taylor & Francis Group, LLC

ISBN: 978-1-032-42842-0 (hbk)
ISBN: 978-1-032-45615-7 (pbk)
ISBN: 978-1-003-37785-6 (ebk)

DOI: 10.1201/9781003377856

Typeset in Times
by Apex CoVantage, LLC

Contents

Preface

This book is intended to serve as an introduction to the area of sustainable and circular economy principles and its application from a process engineering point of view and within the sub-Saharan African context. It is intended for professionals and students in the fields of science and engineering specializing in chemical engineering, civil engineering, mining and metallurgical engineering, environmental chemistry, environmental science, and environmental management.

In line with the sustainable development goals (SDGs) set by the United Nations, Goal 12 speaks to ensuring sustainable consumption and production patterns. Goals 12.5 and 12.2, respectively, state: "By 2030, substantially reduce waste generation through prevention, reduction, recycling, and reuse" and "By 2030, achieve sustainable management and efficient use of natural resources". In order for this to be achieved, there is a need to close the knowledge gap as far as how the principles of circular economy can be successfully implemented in developing countries. African economies are still resource-driven, and with the dwindling of virgin reserves in this continent, there is a need to ensure sustainable use of natural resources.

The book is unique in that it focuses on issues of sustainability and the circular economy within the sub-Saharan Africa (SSA) region by addressing the five loops of circularity – that is, renewable raw materials, product reuse, mechanical recycling, chemical recycling, energy recovery, and carbon utilization. The SSA region is unique in the sense that it is characterized by intensive mining dating back to the nineteenth century, which has led to the accumulation of vast reserves of industrial or mining by-products that can be used as secondary resources. Agriculture and mining are the dominant economic backbones of most African countries, especially SSA, accounting for about 14–30% of the GDP in these countries. Thus, efficient management of waste resources to achieve a circular flow of materials in these sectors will go a long way in preventing environmental degradation, maximizing sustainable use of natural resources, and increasing the GDP contribution from these sectors. The transformation of agriculture in SSA by implementing the five loops of circularity presents a great opportunity for meeting the UN's SDGs.

The primary objective of the book is to enhance the understanding of sustainability and circularity for engineers, scientists, and policymakers and its relevance and application within the sub-Saharan African context. This book covers topics related to the circular economy and the sustainable use of resources in the mining, agricultural, and waste management sectors. It covers fundamental aspects of environmental engineering with an emphasis on unit operations and processes applied in the sustainable management of waste, waste management approaches, and fundamentals of environmental legislation. The legislative framework in various countries grouped by geographical area is presented and evaluated in terms of its effectiveness in addressing the requirements of a sustainable circular economy. The current status of waste management in the mining, agricultural, and municipal sectors in SSA is reviewed, including current challenges and suggested interventions to close the material flow loop. International best practices in sustainability and circularity are presented and used as a benchmark to evaluate the status of the implementation of sustainable development in SSA.

The book includes a number of case studies, discussion questions, and problems, and thus it is adapted to be used as a textbook for undergraduate and postgraduate courses in environmental engineering, environmental science, and environmental engineering. It is also useful for researchers and policymakers working in the area of sustainability within SSA to identify opportunities for further research and development.

Freeman Ntuli

About the Authors

Freeman Ntuli is a professor in the Department of Chemical, Materials, and Metallurgical Engineering at the Botswana International University of Science and Technology (BIUST) and a visiting professor at the University of Johannesburg. He holds a PhD in chemical engineering from the University of Cape Town and has about 20 years of experience working in academia, industry, and government regulatory agencies. His main areas of research are in environmental process engineering and hydrometallurgy. He has published extensively in the areas of waste valorization, industrial wastewater treatment, and circular and sustainable chemical engineering. Professor Ntuli is rated as an established researcher by the National Research Foundation (South Africa) and is a chartered member of the Chartered Institution of Water and Environmental Management (CIWEM) and a chartered scientist (CSci).

Tebogo Mashifana is an educator and researcher specializing in circular economy research. With a wealth of experience in the field, Mashifana holds a DPhil in engineering and has made significant contributions to sustainable resource management. Her expertise lies in waste reduction strategies, green economy, resource efficiency and recovery, and the implementation of circular economy models across various sectors. As an accomplished author, Mashifana has published numerous journal articles, offering valuable insights into the intersection of environmental engineering and circular economy principles. Mashifana's dedication to mentoring students and driving interdisciplinary collaboration underscores her commitment to advancing sustainable development practices. Through her work, Mashifana continues to inspire positive change and promote a more sustainable future.

Thandiwe Sithole is an associate professor in the Department of Chemical Engineering, within the Faculty of Engineering and the Built Environment. Her main area of interest is environmental engineering, especially the circular economy and how to use secondary resources to treat wastewater and turn industrial solid waste into useful building and construction materials, including recycling materials and finding new uses for old ones. The main goal of her research is to develop zero-waste discharge models and value-added products from waste material for use in different building and construction applications. Her research areas are driven by the need to attain or achieve the United Nations' Sustainable Developmental Goals. Dr Sithole is an active member of technical committees, contributing to the advancement of engineering, the circular economy, societal impact, and SDGs.

Thabo Falayi holds a BSc (Hons) in applied chemistry (first class) from the National University of Science and Technology in Zimbabwe, an MTech in chemical engineering technology, and a DPhil in civil engineering from the University of Johannesburg. Thabo has over 10 years of industrial experience as a product development chemist and sampling and water treatment chemist, where he oversaw analytical method development and validation together with sampling method validation. He is a fellow of the African Science Leadership Programme. He has published 42 peer-reviewed research articles in environmental engineering with a Google Scholar H index of 10 and an i10 index of 11 with over 397 citations. He has lectured on chemical engineering at the University of Johannesburg, Malawi University of Science and Technology, and UNISA. He is currently a Lecturer in Chemical Engineering within the Department of Mining, Civil, and Process Engineering at the Namibia University of Science and Technology, and also serves as a Senior Research Associate at the University of Johannesburg. He is also a trained mentor in innovation, with a particular emphasis on human-centred design, agile project management, and systems design approach.

1 Overview of Waste Management

Freeman Ntuli, Tebogo Mashifana, and Thandiwe Sithole

1.1 INTRODUCTION

The term "circular economy" (CE) has gained traction among researchers, industrialists, engineers, scientists, and economists since it was first proposed by Pearce and Turner in 1990. Since the year 2000, it has been used as the main driver of achieving sustainable development in many countries and industries. While there can be no single definition in the literature that completely captures the concept of CE, its core principles are based on the 4Rs of waste management (reduce, reuse, recycle, and recover) and the three core principles of sustainability (economy, society, and environment). A holistic definition of CE was proposed by Kirchherr et al. (2017):

> A circular economy describes an economic system that is based on business models which replace the "end-of-life" concept with reducing, alternatively reusing, recycling and recovering materials in production/distribution and consumption processes, thus operating at the micro-level (products, companies, consumers), meso level (eco-industrial parks) and macro-level (city, region, nation and beyond), to accomplish sustainable development, which implies creating environmental quality, economic prosperity and social equity, to the benefit of current and future generations.
>
> **(Kirchherr et al., 2017)**

Thus, CE represents a shift from a linear model of material flow to a circular model. The linear material flow model has been the traditional approach to waste management, and the ultimate disposal by landfilling of unwanted or waste material has been commonly practiced. A typical comparison of the circular and linear economy is shown in Figure 1.1. The linear material flow is based on the "cradle to grave" approach, leading to the extensive accumulation of waste over time. This results in a significant loss of valuable material and depletion of natural resources.

The CE approach views waste as a resource and focuses on the maximum utilization of material resources. This has transformed waste treatment and management facilities into resource recovery and recycling facilities. Consequently, this approach presents a closed-loop strategy for waste management and strives towards a "zero-waste model".

This book will approach the concept of the circular economy from a process engineering point of view by considering the five loops of circulating materials and energy into the system, as shown in Figure 1.2. This approach ensures circularity

DOI: 10.1201/9781003377856-1

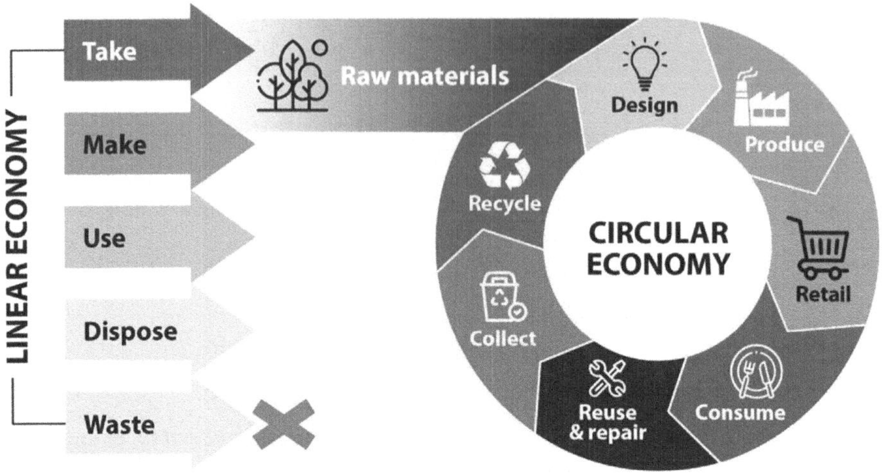

FIGURE 1.1 Comparison of the circular and linear economy.

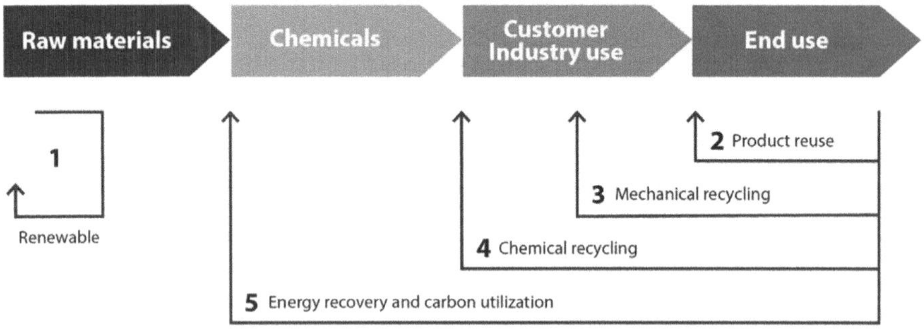

FIGURE 1.2 Circulating materials and energy through the five loops.

Source: Accenture (Elser & Ulbrich, 2017).

is built into the life cycle of a product or materials, preserving their highest utility and value.

1.2 THE SITUATION IN SUB-SAHARAN AFRICA

1.2.1 Geographical Location and Economic Outlook

Sub-Saharan Africa (SSA) is a region that lies south of the Sahara Desert and includes about 48 African countries, excluding the North African region and Sudan (Figure 1.3).

It has a population of approximately 1.18 billion, with an estimated working-age population of 83%, based on the 2021 World Bank estimate. The combined GDP per

FIGURE 1.3 Map of Africa showing Sub-Saharan African countries.

capita in US$ for the region, according to the 2021 World Bank estimate, was 1,626.3 and considered the lowest compared to other regions in the world (see Table 1.1).

The SSA region holds some of the largest natural resources in the world, although it has the lowest GDP per capita. The dominant economic sectors are agriculture, mining, oil, and services. The region is currently considered the least industrialized in the world, with South Africa and Nigeria identified as the largest economies within it. The projected economic growth outlook for the region is extremely uncertain and is being hindered by food insecurity, poor infrastructure, increased inflation and rising global interest rates, rising debt, access to energy and water, and the impact of climate change (Africa, 2023; Ngoran et al., 2016). SSA is considered the most climate-vulnerable region in the world, although it has made an insignificant contribution to global emissions (Africa, 2023). This has significantly impacted the agricultural sector and exacerbated the food insecurity situation. Given the regional endowments in natural resources and renewable energy sources, there

TABLE 1.1
GDP by Region 2021 Estimate

Region	GDP per capita US$
North America	68,369.66
Euro area	42,450.15
EU	38,411.06
Europe and Central Asia	27,152.47
East Asia and Pacific	13,041.78
Latin America and the Caribbean	8,327.61
Arab World	6,226.36
South Asia	2,149.82
Sub-Saharan Africa	1,626.28

Source: Statistica, 2023

are great prospects for investments in sustainable and circular developments that will help address most of the challenges in the region.

1.2.2 Waste Management and Minimization Strategies

As a result of the availability of vast amounts of unutilized land in SSA, which makes landfilling cheaper, a linear approach to the management of waste has been dominant. As a result, vast amounts of municipal solid waste (MSW) and mineral waste are discarded or stockpiled without further processing. A certain level of adherence to the circular economy is evident in the agriculture sector, where agricultural residues are either composted, thereby recycling nutrients back into the soil, or utilized as livestock feed, representing re-use. However, disposal by burning some of the dry agricultural residues is still also commonly practiced.

SSA is also a dominant mining region known for its production and reserves of gold, iron, coal, uranium, diamond, chrome, base, and precious metals. Large-scale mining in this region began in the mid-nineteenth century and has grown over the years into a multi-billion industry. However, the long legacy of mining has also resulted in the accumulation of large stockpiles of mine tailings with their associated environmental challenges. Not only does this occupy large tracts of land that could be used for other economic activities, but it also represents a vast reserve of valuable materials that can be recycled back into the economy. Thus, there is tremendous potential to implement the circular economy approach to drive sustainable development in this region.

Municipal solid waste management (MSWM) in SSA is predominantly dominated by landfilling and open dumping of waste. Waste collection and disposal are centralized and well managed in formal settlements in the urban areas. However, in informal settlements there is no centralized and well-coordinated system for the collection and disposal of waste, resulting in open dumping and burning of waste being predominant. The average waste collection rate in Africa is only about 55% (United Nations Environment Programme, 2018), which shows that the bulk of the MSW generated is not adequately collected and disposed of. Waste segregation and recycling are not widely practiced; however, informal waste recyclers play a critical role in most urban centers in scavenging paper, plastic, and metals for resale to buy-back centres. Overall, although 70–80% of the MSW generated in Africa is recyclable, the recycling rate is only about 4 (United Nations Environment Programme, 2018). Thus, there are great economic benefits to be derived from the implementation of CE in the waste management sector in Africa. However, the current challenges of waste collection and management, inadequate legislation, and expertise need to be addressed if the circular economic model is to be effectively implemented. If properly implemented for MSWM, CE can inject an estimated US$8 billion/year into African economies, which can greatly alleviate poverty and unemployment and improve the quality of the environment.

1.3 SUSTAINABLE DEVELOPMENT AND ENVIRONMENTAL SUSTAINABILITY

1.3.1 Introduction

The waste management market is growing drastically in Africa. According to Mordor Intelligence, approximately 4% of the waste generated in Africa, which includes municipal solid waste, industrial waste, electronic waste, hazardous waste, biomedical waste, and plastic waste, is recycled. Most of the waste, about 90% is disposed of in controlled or uncontrolled landfills. One of the African Union's visions is for Africa to recycle at least 50% of the waste generated in urban areas by 2023.

The *Africa Waste Management Outlook 2018* has reported the status quo of waste management in the African continent, and the following predictions are presented:

- The total municipal solid waste generated will rise to 500×10^3 tons/year in African countries by 2025.
- Africa is a dumping ground for many countries around the globe, with 19 of the 50 largest dumpsites located in Africa.

Waste management is critical for economic, social, and environmental sustainability. Currently, there are poor waste management systems in Africa, which result in environmental pollution and soil and water contamination, destroying the ecosystem and affecting human health. To achieve SDGs regarding waste management, any nation must develop and implement a waste management strategy. The concept of sustainable development focuses on the current environment, society, and economy without compromising the future of the next generation. A circular economy is a key concept for sustainable waste management. The common practice in dealing with solid waste in Africa is through a linear approach, whereby raw materials are introduced into systems, products are developed, and waste is generated. Depending on the type of waste generated, the generated waste is then disposed of in the environment at high costs. A sustainable solution to waste management would include a closed-loop system or a circular system that incorporates 4R (Reduce, Reuse, Recycle, and Recover) strategies. A linear approach to dealing with and handling waste has other implications and is not environmentally sustainable.

The United Nations' SDGs for 2030 and SDG 12 are a blueprint for responsible consumption and production. Over the years, there has been a drastic increase in the human footprint. Reducing the human footprint is key to sustainable development.

1.3.2 Sustainable Mining Waste Management

Africa is a continent with key mineral commodities such as diamond, gold, coal, iron ore, cobalt, copper, and bauxite. The major mining activities in Africa are conducted in the Democratic Republic of Congo, Namibia, South Africa, and Zimbabwe. With the mining activities in Africa, the generation of mining waste is inevitable. Figure 1.4 shows the top countries in Africa among the top ten global resource countries.

There is no doubt that the mining process generates significant quantities of waste, which requires adequate treatment to achieve environmental sustainability (Aznar-Sánchez et al., 2018). The common mining waste is in the form of tailings, slurry, waste rock, and so on. The waste is usually stored in tailings dams that occupy large areas and pollute the environment. Most of the mining waste is hazardous and contains contaminants that pollute the water, soil, and air. Besides affecting human health, contamination of water has a direct impact on food security

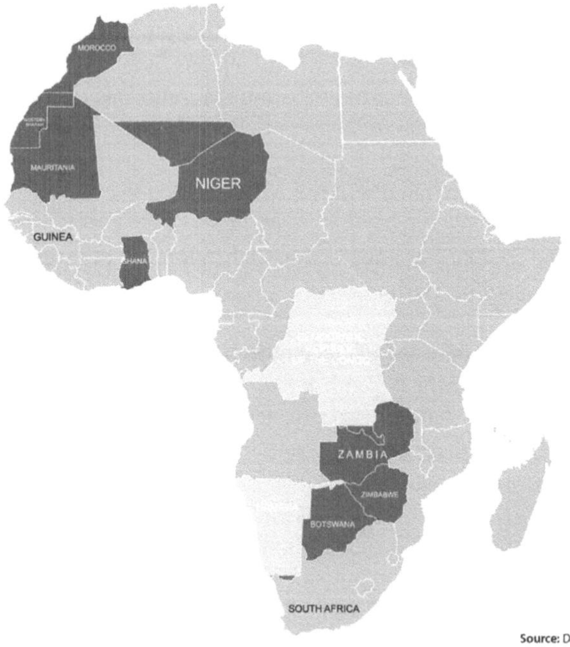

	Resources	Share of global resources, %
North Africa	Phosphate	32
West Africa	Bauxite	40
	Uranium	5
	Iron ore	4
Central and Southern Africa	Platinum	88
	Chromium	84
	Diamonds	60
	Cobalt	49
	Gold	40
	Uranium	13
	Copper	5

Source: Deutsche Bank, US Geological Survey (USGS). McKinsey Quarterly, World Nuclear Association, U.S. Global Investors

FIGURE 1.4 Top African countries among the top ten global resource countries.

due to its effect on domestic animals and aquatic life. Waste should be minimized, reused, and recycled to achieve environmental sustainability. The reprocessing of mining waste has increased in recent years to recover valuable constituents and utilize them as raw materials for various applications, including construction and building. This approach to handling and managing mine wastes is critical in preventing environmental pollution while contributing to SDGs. Tayebi-Khorami et al. (2019) proposed an integrated approach for handling mine waste, as shown in Figure 1.5.

The four pillars for achieving sustainability are human, economic, environmental, and societal. The proposed approach incorporates these four pillars. Through the integrative approach presented, the social dimensions are one of the areas for the mining industry to minimize its social impact. Mining companies must develop and implement technologies that do not harm the society and the environment they operate in order to prevent social conflicts that may affect the economy. It is imperative that the geo-environmental aspects of the mining industry are considered in order to prevent any pollution and are prioritized to achieve the circular economy aspiration. For sustainable development, the mining industry should expand its geo-metallurgical processes to incorporate environmental and economic information (Bridge et al., 2013).

The International Council on Mining and Metals (ICMM) has presented 10 principles for effective sustainable development (see Figure 1.6).

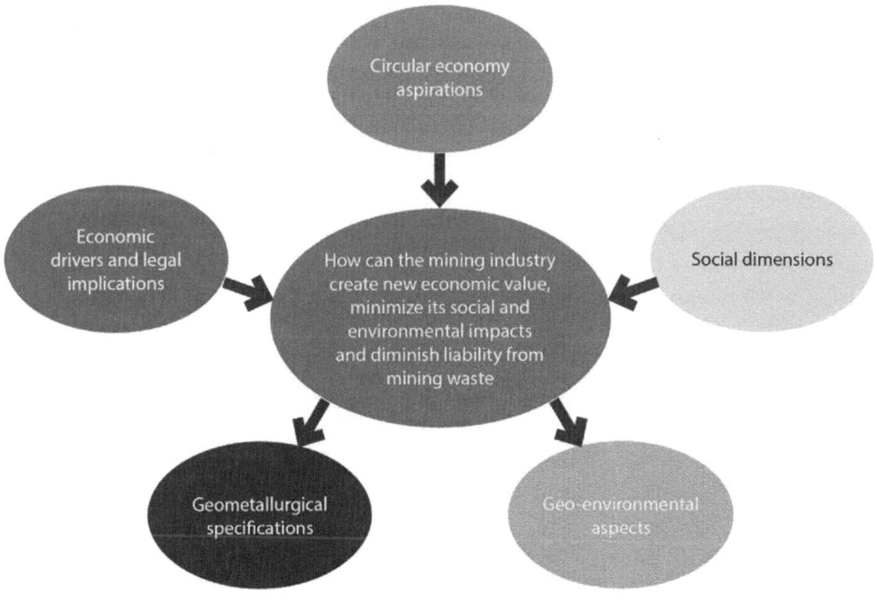

FIGURE 1.5 Key areas of an integrative approach for mining waste.

Source: Tayebi-Khorami et al., 2019.

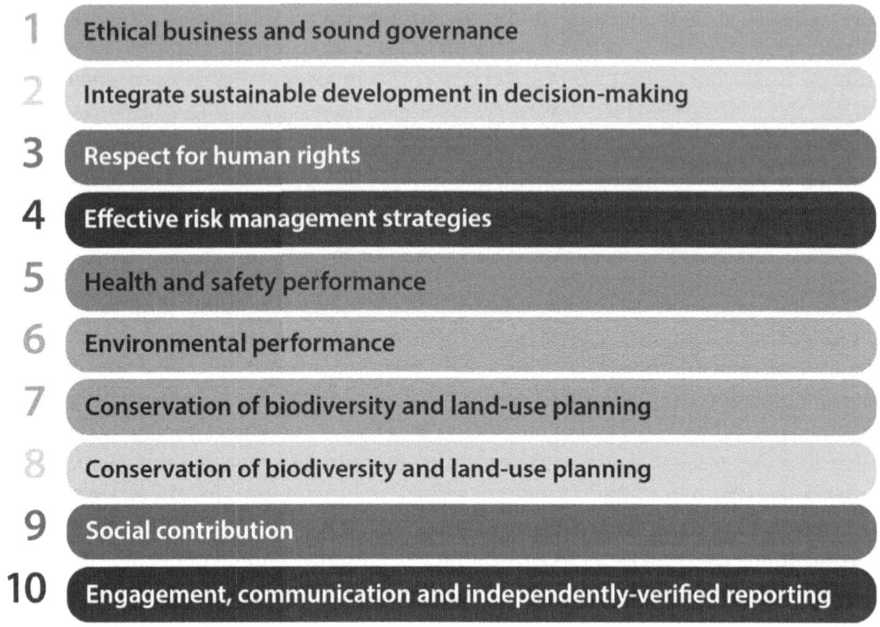

FIGURE 1.6 Ten principles for effective sustainable development of ICMM.

1.3.3 The Need for Sustainable Waste Management

To sustainably manage waste, the collection, transportation, valuing, and disposal of different types of waste must not harm the environment, human health, or future generations. Although waste management through collecting, transporting, and disposing of waste in landfills is not the most sustainable alternative and is being discouraged in the world today, it is still very prevalent. In order to achieve zero waste as per the SDGs plan, new concepts such as circular economies and improving environmental sustainability are emerging and being emphasized (Sithole et al., 2022). In a circular economy, sustainable waste management plays a crucial role that is beneficial to society and the economy. The collection, treatment, reuse, and recycling through the valorization of waste constitute sustainable waste management and, if properly employed, can provide energy and resources, such as value-added products (Abdel-Shafy & Mansour, 2018). The reduction of waste materials through valorization is the main goal of the circular economy, which is more beneficial to the environment (reuse of waste and lower emissions of pollutants). Sustainability refers to reducing the number of natural resources consumed, repurposing all natural resources as much as possible, producing as little waste as possible, or valorizing the waste. Environmental sustainability is commonly understood within the context of sustainable development, a model that addresses the conflicting goals of economic growth, environmental protection, and social development (Manno & Fix, 2010). A proper valorization of waste and management is a suitable alternative for job creation, improvement of waste management methods, minimizing the environmental impact, and maintaining a healthy environment, a direct response to improve environmental sustainability. Waste management through valorization includes the creation of energy through landfill gas, the development of value-added products such as building and construction materials, and the production of fuel from plastic waste, among others.

1.3.4 Municipal Solid Waste Management

The term "municipal solid waste" includes a wide variety of materials collected by the municipal or disposed of at the municipal waste disposal site, including materials generated by residential, industrial, institutional, commercial, municipal, and construction and demolition activities (Hoornweg et al., 2015). Municipal waste is a combination of household, commercial, and demolition waste; electronic waste, which includes computer components; and radioactive waste. In many African countries, rapid urbanization, industrialization, and changes in consumption patterns have contributed to the growing problem of MSW. Statistics show that in 2012 there were 125 million tonnes of municipal solid waste generated across Africa, of which 65% were from sub-Saharan Africa (Godfrey et al., 2019). Figure 1.7 shows the projected rate of increase in MSW generation, which is envisaged to increase to 244 million tonnes by the year 2025 and 368 million tonnes by 2040, which is a threat to the health of the public and the environment.

Available data shows that in 2012 only 54% of the waste was collected in Africa, and it is anticipated that there will be an increase in MSW collected by the year 2025 (Hoornweg & Bhada-Tata, 2012). This is due to poor funding, inadequate legislation and policies, and ineffective awareness campaigns that pose challenges to MSW

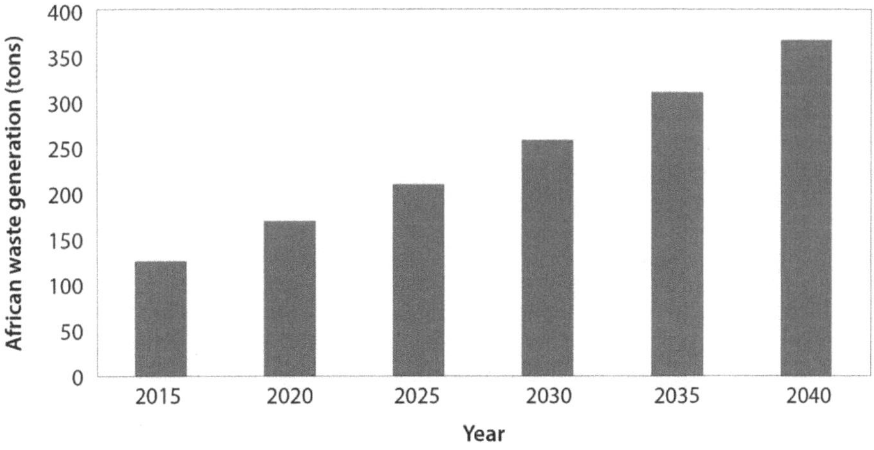

FIGURE 1.7 Municipal waste generation in Africa.

Source of data: UNEP. (2018). *Africa Waste Management Outlook* (p. 15). UN Environment Programme.

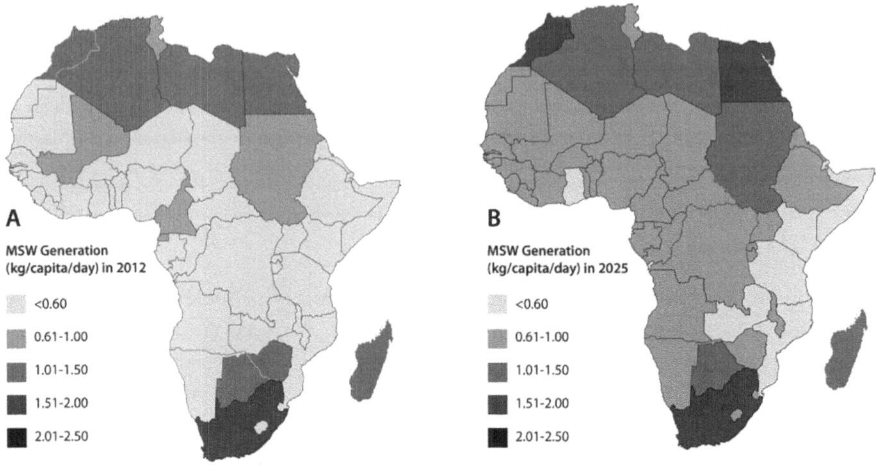

FIGURE 1.8 African countries' average daily waste generation in 2012 (A) and in 2025 (B) by geographic region.

Source: United Nations Environment Programme, 2018; Godfrey et al., 2019.

management in SSA. Compared to the global average of 1.24 kg per day, Africa generated 0.78 kg per capita waste in 2012 (Scarlat et al., 2015). However, there are considerable spatial differences in the amount of waste generated (Figure 1.8), which range from as low as 0.09 kg per day (Ghana) to as high as 2.98 kg per day (Seychelles). It is common for Small Island Developing States to generate high amounts of waste per capita due to the high levels of tourism and better waste accounting (Hoornweg & Bhada-Tata, 2012). Therefore, there is a need to implement strategies to manage this waste not only effectively but also sustainably. To sustainably manage waste, the collection, transportation, valuing, and disposal of MSW must not threaten the health of the public and the environment. Therefore, proper management and monitoring of various waste materials are essential.

1.4 OVERVIEW OF WASTE AS A RESOURCE IN SUB-SAHARAN AFRICA

1.4.1 Introduction

The dictates of a circular economy demand that resources be kept in the economic loop for as long as possible. This then allows for a reduction in the need for virgin materials in the economic loop. Therefore, the utilization of waste materials as secondary resources contributes to the sustainable use of natural resources.

1.4.2 Plastic Waste

Besides the plastic waste generated by anthropogenic activities in Africa, there has been a steady importation of plastic waste into Africa. Over 172 million tonnes of plastic waste were imported into Africa in the period 1990–2017 (Babayemi et al., 2019). The highest importers of plastic waste in Africa are Egypt, Nigeria, and South Africa. These three countries are also the highest in terms of plastic production, with South Africa producing 9 million tonnes per year, Nigeria 2.3 million tonnes per year, and Egypt 4 million tonnes per year (EUROMAP, 2016). The most common monomers for these plastics are polyolefins, namely polyvinyl chloride (PVC), polypropylene (PP), polypropylene (PP), and polystyrene (PS), low-density polyethylene (LDPE), and high-density polyethylene (HDPE).

1.4.3 Mining Waste

Mining dominates the economy of most African countries, leading to the production of tailings dams. Tailings, which are waste products in the beneficiation process, are generally stored in a slurry form. It has been reported that there are over 270 gold mine tailings in South Africa covering a space of 180 square kilometres (Falayi et al., 2018). In the Copperbelt of Zambia, over 9125 ha of land is occupied by tailings (Lindahl, 2014). Zimbabwe is said to be sitting on around 500 tailings dams (World Mine Tailings Failures, 2022). There are no comprehensive inter-country studies, making country-tocountry-comparisons challenging.

1.4.4 E-Waste

E-waste has become a waste of concern due to the booming consumption or utilization of electronic equipment like computers and smartphones. This is an emerging waste in Africa with very little data available for generation and disposal. However, it is important to note that the legislation for the control of e-waste in Africa is still in its infancy. Africa is a ready market for used working electronic equipment, and as such e-waste is likely to increase in Africa. On average the production of e-waste is about 2 kg/inhabitant/year in Africa, with the highest of 12 kg/inhabitant/year in Botswana and Tunisia (Mihai et al., 2019).

CASE STUDY 1: PLASTIC POLLUTION A SIGNIFICANT DRIVER OF CLIMATE CHANGE AND BIODIVERSITY LOSS – NEW REPORT (ADAPTED FROM HTTPS://WWW.CIWEM.ORG/NEWS/CONNECTING-THE-DOTS-PLASTIC-POLLUTION-REPORT-ENVIRONMENTAL-INVESTIGATION-AGENCY)

The problem of waste plastic has become a global concern. The Environmental Investigation Agency (EIA), an organization that "investigates and campaigns against environmental crime and abuse has labelled plastic pollution as the most prevalent environmental pollutant and as such "a significant driver of climate change and biodiversity loss". Based on current projections, it is expected that by 2050 the weight of plastic will likely far exceed the weight of all fish in every ocean on earth, leading to significant loss of aquatic life. Furthermore, the production of plastics has a significant impact on the environment, with 99% of plastics made from fossil fuel feedstocks, and the process of production releases about 1.89 tonnes of carbon dioxide equivalent (CO_2e) per tonne of virgin (new) plastic produced.

The report recommends the following to avert plastic pollution:

- Eliminate the discharge of plastics into the environment
- Phase plastic production down to sustainable levels in alignment with Sustainable Development Goal (SDG) 12
- Promote the upscaling of reuse, refill, and traditional packaging systems tailored to national contexts.

Ayeleru et al. (2020) published a review on the challenges of plastic waste generation and management in sub-Saharan Africa. Based on this article and the background information given above, together with other relevant resources:

1. What proportion of waste generated in sub-Saharan Africa does plastic waste (PW) account for?
2. What are some of the challenges in the waste management systems in sub-Saharan Africa that exacerbate the problem of PW?
3. How can the principles of the circular economy and sustainability be used to address the challenges of PW in sub-Saharan Africa?

Based on Figure 1.2, how can the five loops circulating materials and energy be implemented for managing the production and use of plastics in sub-Saharan Africa?

CASE STUDY 2: SUSTAINABLE DEVELOPMENT GOALS

The United Nations agreed upon the SDGs in 2015, with a target completion date of 2030. The SDGs are a set of goals designed to address the environmental, socioeconomic, and political challenges confronting our world. A pictorial presentation of the SDGs is shown in Figure 1.9.[1]

FIGURE 1.9 Illustration of the Sustainable Development Goals.

A review of the progress made towards achieving the SDGs midway through the implementation period in 2023 has revealed that the world is not on track to achieve the 17 SDGs and urgent action is needed by global leaders to remove the roadblocks and accelerate the progress (Malekpour et al., 2023). Of the 36 targets they reviewed, only 2 were on track as of 2023, namely, access to mobile networks and internet usage, while 14 showed fair progress, 12 showed limited or no progress, and 8 targets were identified as deteriorating. Some of the challenges hampering progress were identified as global events such as the COVID-19 pandemic and the war in Ukraine. However, the major impediments to progress were identified as:

- Inaction by governments
- Lack of financial support and inadequate administrative structures
- A culture of consumerism driven by the advertising industry making it difficult for consumers to shift their behaviour towards sustainable diets, transport, and consumption patterns
- Existing investments in fossil-fuel infrastructure have generated resistance to the just energy transition.

Malekpour et al. (2023) identified three priority areas that scientists and policymakers can focus on to help accelerate progress: the identification and a tailored approach to remove key roadblocks to progress, finding feasible and cost-effective transformation pathways, and strengthening governance and accountability.

By focusing on the following SDG goals: 6 (clean water and sanitation), goal 7 (affordable and clean energy), goal 12 (responsible consumption and production), and goal 13 (climate action) within the SSA context:

1. What are the key roadblocks to achieving progress?
2. What are the feasible and cost-effective transformation pathways that can be used to remove the key roadblocks?
3. What governance and accountability structures can be put in place to ensure the specified SDG goals are met?

1.5 PROBLEMS

1. What are the four main pillars of sustainable development? How is each of these pillars an enabler or a hindrance to sustainable development within the sub-Saharan context?
2. Identify the core principles on which the concept of the circular economy was built.
3. Compare and contrast the linear economy against the circular economy.
4. How can the UN's SDGs enhance the transition towards a circular economy in sub-Saharan Africa?
5. Identify the major waste management challenges in sub-Saharan Africa.

NOTE

1 https://www.un.org/sustainabledevelopment/news/communications-material/

REFERENCES

Abdel-Shafy, H. I., & Mansour, M. S. (2018). Solid waste issue: Sources, composition, disposal, recycling, and valorization. *Egyptian Journal of Petroleum*, *27*(4), 1275–1290.

Africa, S. (2023). *Regional economic outlook, Sub-Saharan Africa, October 2023*. https://doi.org/10.5089/9798400256806.086

Ayeleru, O. O., Dlova, S., Akinribide, O. J., Ntuli, F., Kupolati, W. K., Marina, P. F., Blencowe, A., & Olubambi, P. A. (2020). Challenges of plastic waste generation and management in sub-Saharan Africa: A review. *Waste Management*, *110*, 24–42.

Aznar-Sánchez, J. A., García-Gómez, J. J., Velasco-Muñoz, J. F., & Carretero-Gómez, A. (2018). Mining waste and its sustainable management: Advances in worldwide research. *Minerals*, *8*(7), 1–27.

Babayemi, J. O., Nnorom, I. C., Osibanjo, O., & Weber, R. (2019). Ensuring sustainability in plastics use in Africa: Consumption, waste generation, and projections. *Environmental Sciences Europe*, *31*(1), 60.

Bridge, R., Brosig, D., Lozano, C., & Laurila, H. (2013). *Geometallurgy: An underutilised technology*. Canadian Institute of Mining, Metallurgy and Petroleum.

Elser, B., & Ulbrich, M. (2017). *Taking the European chemical industry into the circular economy*. Accenture.

EUROMAP. (2016). *Plastics resin production and consumption in 63 countries worldwide*. Retrieved July 1, 2020, from http://www.pagder.org/images/files/euromappreview.pdf

Falayi, T., Ntuli, F., & Okonta, F. N. (2018). Synthesis of a paste backfill geopolymer using pure acidic gold mine tailings. *Journal of Solid Waste Technology and Management, 44*, 311–320.

Godfrey, L., Ahmed, M. T., Gebremedhin, K. G., Katima, J. H., Oelofse, S., Osibanjo, O., Richter, U. H., & Yonli, A. H. (2019). Solid waste management in Africa: Governance failure or development opportunity. *Regional Development in Africa*, 235.

Hoornweg, D., & Bhada-Tata, P. (2012). *What a waste: A global review of solid waste management* (Urban Development Series; Knowledge Papers No. 15). World Bank.

Hoornweg, D., Bhada-Tata, P., & Kennedy, C. (2015). Peak waste: When is it likely to occur? *Journal of Industrial Ecology, 19*(1), 117–128.

Kirchherr, J., Reike, D., & Hekkert, M. (2017). Conceptualizing the circular economy: An analysis of 114 definitions. *Resources, Conservation and Recycling, 127*, 221–232. https://doi.org/10.1016/j.resconrec.2017.09.005

Lindahl, J. (2014). *Environmental impacts of mining in Zambia: Towards better environmental management and sustainable exploitation of mineral resources* (SGU No. 2014:22). Geological Survey of Sweden. Retrieved February 16, 2017, from http://www.meetingpoints-mining.net/wp-content/uploads/2014/06/Environmental-impacts-of-mining-in-Zambia.pdf

Malekpour, S., Allen, C., Sagar, A., Scholz, I., Persson, Å., Miranda, J. J., Bennich, T., Dube, O. P., Kanie, N., Madise, N., Shackell, N., Montoya, J. C., Pan, J., Hathie, I., Bobylev, S. N., Agard, J., & Al-Ghanim, K. (2023). What scientists need to do to accelerate progress on the SDGs. *Nature, 621*(September), 250–254. https://doi: https://doi.org/10.1038/d41586-023-02808-x

Manno, J., & Fix, A. (2010). Environmental sustainability and sustainable development. In *Oxford research encyclopedia of international studies*. Oxford University Press.

Mihai, F. C., Gnoni, M. G., Meidiana, C., Ezeah, C., & Elia, V. (2019). Waste electrical and electronic equipment (WEEE): Flows, quantities, and management—a global. *Electronic Waste Management and Treatment Technology*. https://doi.org/10.1016/B978-0-12-816190-6.00001-7

Mordor Intelligence. (n.d.). *Africa waste management market—growth, trends, COVID-19 impact, and forecasts (2022–2027)*. Mordor Intelligence.

Ngoran, S. D., Xue, X. Z., & Wesseh, P. K. (2016). Signatures of water resources consumption on sustainable economic growth in Sub-Saharan African countries. *International Journal of Sustainable Built Environment, 5*(1), 114–122.

Pearce, D. W., & Turner, R. K. (1990). *Economics of natural resources and the environment*. The Johns Hopkins University Press.

Scarlat, N., Motola, V., Dallemand, J. F., Monforti-Ferrario, F., & Mofor, L. (2015). Evaluation of energy potential of municipal solid waste from African urban areas. *Renewable and Sustainable Energy Reviews, 50*, 1269–1286.

Sithole, N. T., Tsotetsi, N. T., Mashifana, T., & Sillanpää, M. (2022). Alternative cleaner production of sustainable concrete from waste foundry sand and slag. *Journal of Cleaner Production*, 130399.

Tayebi-Khorami, M., Edraki, M., Corder, G., & Golev, A. (2019). Re-thinking mining waste through an integrative approach led by circular economy aspirations. *Minerals, 9*(5), 286.

United Nations Environment Programme. (2018). *Africa waste management outlook—summary for decision-makers*. Retrieved April 17, 2022, from https://wedocs.unep.org/20.500.11822/25515

World Mine Tailings Failures—from 1915. Retrieved January 10, 2022, from www.worldminetailingsfailures.org/estimate-of-world-tailings-portfolio-2020/

2 Waste Management Principles and Their Application

Tebogo Mashifana

2.1 INTRODUCTION

Waste management principles include the protection of health and the environment, as well as the use of modern techniques such as energy generation from waste, composting, and recycling of biological materials (Gul et al., 2022). Furthermore, these principles involve a transition towards a circular economy, where waste is regarded as a valuable resource, and endeavours are concentrated on reduction, recycling, and treatment techniques (Guman & Wegner-Kozlova, 2020). Within the realm of international legislation, waste management principles encompass the notion of permanent sovereignty over natural resources, preventive measures, collaboration, sustainable development, and the precautionary principle (Rayfuse, 2016). In practical terms, these principles are implemented across diverse industries, with the construction sector being a notable example, wherein material waste poses a significant challenge. The principles of waste management are deployed to diminish material waste, enhance waste management and storage practices, and propagate the concept of reduce, recycle, and reuse (Laovisutthichai & Bao, 2022). Additionally, waste management principles find application within the industrial domain through the adoption of cleaner production technologies and advanced treatment methodologies for hazardous waste.

The management of waste has posed a global challenge due to population growth, human activities in many parts of the world, and industrialization and urbanization leading to various wastes. Developing sustainable systems for waste management is difficult under these circumstances. In Africa, landfilling remains the standard practice for managing waste, while Europe introduced its Waste Framework Directive in 2018, which incorporates a five-step hierarchy system into their policies based on the approach termed "3-R". Landfilling should be used only as a last resort when handling any form of refuse; it is essential that different industries begin implementing both this efficient methodology and utilizing hierarchical approaches towards responsible disposal, such as through prevention/pre-planning or reuse where possible, rather than making direct landfilling (Figures 2.1 and 2.2).

Africa's waste predicament can be addressed by employing a diverse array of waste management models, concepts, and applications. According to the United Nations, it is clear that the existing waste management practices in Africa have

DOI: 10.1201/9781003377856-2

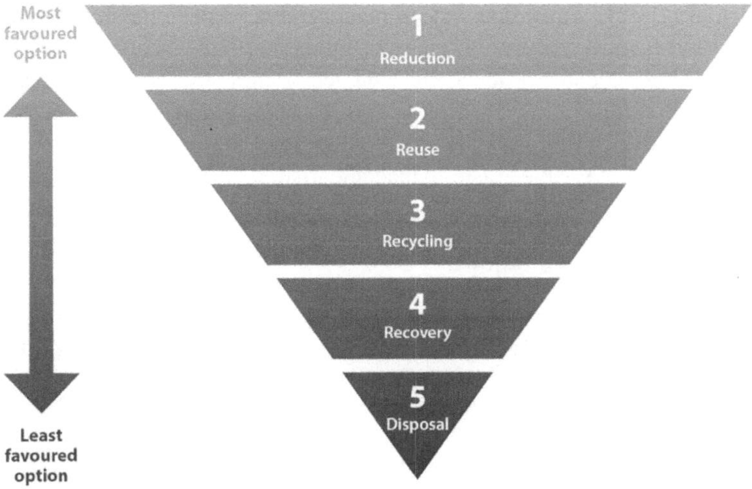

FIGURE 2.1 3-R (Reduce, Reuse, and Recycle).

Source: Li et al., 2023.

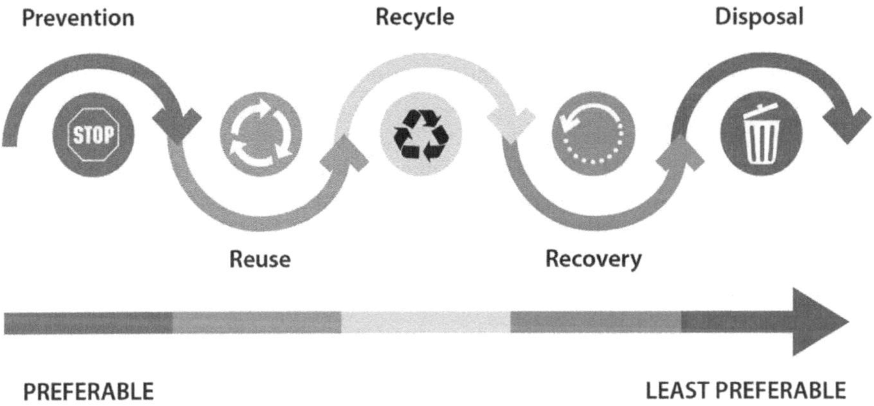

FIGURE 2.2 Five-step waste hierarchy.

detrimental effects on the environment, as well as social and economic implications. The principles of waste management that are applicable in Africa are not limited to other regions of the world. The primary objectives of these principles strive to ensure proper waste management and disposal while mitigating the detrimental impacts on the environment and public health.

In Africa, the following waste management concepts are of utmost importance: recycling of waste and the economic and environmental gains attributed to recycling (Ogutu & Kathambi, 2022). The proper management of electronic and electrical waste, including methods for recovering metals from e-waste, is crucial (Moyen &

Archodoulaki, 2023). Hazardous waste management focuses on source reduction, recycling, and landfill options (Akpan & Olukanni, 2020), including transforming plastic waste into valuable products through eco-friendly techniques, such as utilizing them as a component in cementitious composites in road construction, as well as producing liquid fuels from plastic waste (Amankwa et al., 2021). At both national and local levels, sustainable waste management strategies are implemented, which include comprehensive urban planning, the use of information and communication technologies (ICT), financial mechanisms, community participation, and private sector involvement (Ibrahima, 2019).

2.2 WASTE COLLECTION, STORAGE, AND RESOURCE RECOVERY

The yearly production of waste has been consistently rising, with estimations indicating that it will reach 269 million metric tons by 2030, an increase from the 174 million tons in 2016. Nevertheless, the rates of collection and coverage for MSW have yet to achieve the desired 100%, leading to substantial quantities of uncollected waste in public areas, notably in impoverished regions (Wikurendra et al., 2023).

The incorporation of circular economy principles by enterprises and business owners in Uganda indicates a promising opportunity for waste management in sub-Saharan Africa, resulting in positive consequences for the environment, economy, and society [4]. However, the implementation of circular economy practices in this area is hindered by several obstacles, including the absence of educational initiatives, insufficient political dedication, and inadequate funding and national regulations concerning waste management (Faried et al., 2023). Businesses and entrepreneurs in Uganda embracing circular economy principles provide a promising path for handling waste in sub-Saharan Africa. This approach brings positive results for the environment, economy, and society, offering hope for improved waste management in the region.

Geme et al. (2023) indicate that various stakeholders in Uganda, including the government, private sector, academic community, and civil society, have shown awareness of and enthusiasm for circular economy concepts. Nevertheless, these stakeholders must collaborate and possess a more comprehensive understanding of the situation. Buda and Ricz (2023) underscore the significance of a regulatory and policy framework that takes local specificities into account and incorporates bottom-up initiatives to promote environmentally friendly development. Moreover, Faried et al. (2023) emphasize the importance of empowering communities in waste management to establish a sustainable system. These findings indicate that Uganda can benefit from the application of circular economy ideas by addressing waste management challenges and fostering stakeholder collaboration.

Waste collection in sub-Saharan Africa faces several challenges. One of the main challenges is the continued practice of the old traditional linear waste economy, which involves taking, making, and disposing of waste (Debrah et al., 2022). This linear approach hinders the transition to a circular economy model, which is necessary for sustainable development (Adeniran et al., 2022). Lack of educational programs and public awareness about waste management activities is another significant challenge (Ndam et al., 2023). Furthermore, there is a shortage of political determination, funding, and national policies regarding waste management (Buda, 2022). Improper disposal of plastic waste is a significant concern because it leads to harmful environmental and health consequences (Adedara et al., 2023). Rapid urbanization and increasing poverty in cities like Yaoundé, Cameroon, further complicate household waste management. Inadequate waste collection and coverage rates, especially in low-income neighbourhoods, are also major challenges. These difficulties show the significance of enhancing awareness and developing better waste collection infrastructure and waste management methods in sub-Saharan Africa.

Unregulated waste collection methods can result in ineffective waste management practices and increased levels of litter and pollution. Official waste collection systems, usually run by local governments or private firms, face obstacles like inadequate infrastructure, funding shortages, and inadequate waste management regulations. In some cases, waste collection vehicles are in poor condition, and collection routes may be poorly designed or lack proper signage.

To enhance waste collection in sub-Saharan Africa, it is imperative to allocate resources towards vital infrastructure, assets, and effective waste management regulations. This necessitates the construction of novel landfill sites, the modernization of waste collection vehicles, and the establishment of formal waste collection systems in underserved regions. It is critical for governments, non-governmental organizations (NGOs), and other stakeholders to work together to support recycling, reduce trash, and increase public understanding of appropriate waste management techniques. To tackle these problems, a thorough approach that accounts for the specific social, economic, and environmental conditions of sub-Saharan African nations is required. By collectively augmenting waste collection and management practices, we can foster environmental sustainability, diminish pollution, and fortify public health within sub-Saharan African communities. Figure 2.3 shows the conceptual framework: the relationship of waste collection systems and resource recovery.

The Property-Close Collection System, also known as Kerbside or Door-to-Door collection, involves waste being picked up directly from outside homes or buildings. Residents place their trash in bins or bags, which are then collected by a truck. Alternatively, the Drop-Off Collection System requires individuals to take their trash to designated locations where waste from the community is gathered. These sites typically consist of large bins or areas where various types of waste can be deposited. Another approach is the Buy-Back Collection System, where individuals can bring certain items such as bottles or cans to designated centres and receive payment for them. This system incentivizes recycling by providing monetary rewards for returning reusable items. Lastly, the Deposit-Refund Collection System, also known as the Returnable Container Legislation system, involves consumers paying a small deposit when purchasing items like bottled drinks. Upon returning the empty container to the store, consumers receive a refund of their deposit (Mwanza et al., 2018).

2.2.1 Waste Storage

Waste storage in sub-Saharan Africa faces several challenges related to inadequate infrastructure and resources, limited access to waste management services, and high rates of urbanization (Ossei-Bremang et al., 2023). In many sub-Saharan African countries, waste is stored in open dumps or in unlined landfills, which can lead to

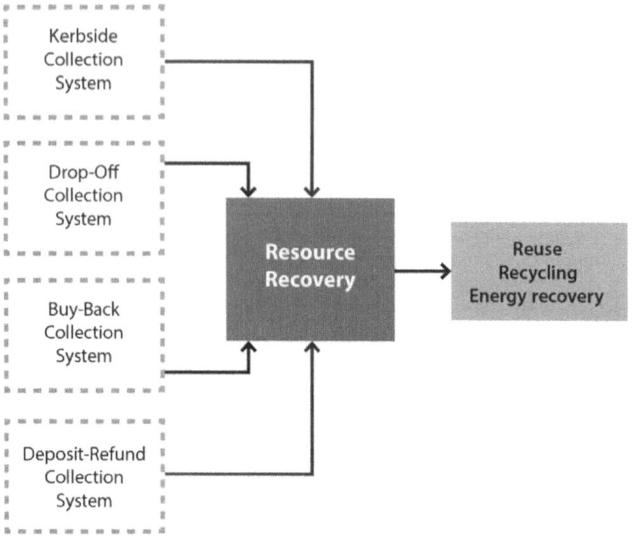

FIGURE 2.3 Conceptual framework: Relationship of waste collection systems and resource recovery.

Source: Mwanza et al., 2018.

groundwater contamination and soil pollution. Large-scale agricultural waste generation in the region poses significant challenges to its disposal (Apazhev et al., 2022). Additionally, the storage of waste-based briquettes made from oil palm shells and decanter cakes can lead to a reduction in efficiency over time (Ossei-Bremang et al., 2023). The waste management systems in sub-Saharan Africa are not comprehensive, and there is a need for better waste management procedures and environmental control policies (Idowu et al., 2019).

Furthermore, waste trade involving the transboundary shipment of toxic waste from developed to developing countries is a major concern in sub-Saharan Africa, with weak regulatory frameworks allowing for reckless dumping of hazardous waste (Duodu & Mpuure, 2023). Overall, there is a need for improved waste storage and management practices in sub-Saharan Africa to address environmental and public health issues. One of the challenges of waste storage in sub-Saharan Africa is the limited availability of adequate storage facilities. Many areas lack proper waste storage containers, and waste may be stored in bags or open bins, which can attract pests and animals and create unsanitary conditions.

Another challenge is the lack of proper waste segregation and disposal practices. In many sub-Saharan African countries, waste is not properly segregated at the source, which can make it difficult to manage and recycle. This can also lead to the accumulation of hazardous waste materials, such as electronic waste or medical waste, which require special handling and disposal methods.

To improve waste storage in sub-Saharan Africa, it is important to invest in adequate infrastructure and resources and establish formal waste management systems in areas that lack access to these services. This can include building new landfill sites, upgrading waste storage facilities, and establishing waste segregation and disposal practices. Governments, NGOs, and other stakeholders can also work together to promote waste reduction and recycling and raise awareness about the importance of responsible waste management practices.

In addition to these efforts, it is important to address the root causes of waste generation in sub-Saharan Africa, such as poverty, inadequate housing, and limited access to basic services.

By promoting sustainable development and addressing these underlying factors, it is possible to reduce waste generation and improve waste storage and management practices in sub-Saharan African communities.

2.2.2 Waste Recovery

Waste resource recovery in sub-Saharan Africa is an emerging area that involves the recovery of valuable materials from waste, such as recyclable materials, organic matter, and energy. This approach offers several benefits, including reducing waste generation, creating jobs, and promoting environmental sustainability. Waste resource recovery refers to turning waste into usable resources by means of procedures including composting, recycling, and the production of biofuel. These processes offer alternative fuel sources and job opportunities for marginalized communities, particularly women and youth (Agbabiaka et al., 2022). The management of institutional waste in sub-Saharan Africa is similar to MSW management, but methods like landfill and incineration contribute to environmental challenges and greenhouse gas emissions (Ddiba, 2020). However, source sorting and the development of semi-automated material recovery facilities can minimize emissions and improve waste management (Njenga & Mendum, 2018). In urban areas, households can contribute financially to support sustainable solid waste management, and solid waste recovery alternatives can turn waste into valuable resources, save landfill space, and create employment opportunities (Akinjogbin et al., 2020). In general, waste resource recovery in sub-Saharan Africa holds promise for addressing environmental issues, yielding financial gains, and advancing sustainable development objectives (Mombo & Bigirwa, 2017).

In many sub-Saharan African countries, waste resource recovery is carried out by informal waste pickers, who collect and sort waste for recycling or sale. This informal sector not only provides an important source of livelihood for many people but also faces several challenges related to inadequate infrastructure and resources, lack of access to proper equipment and protective gear, and unsafe working conditions.

To promote waste resource recovery in sub-Saharan Africa, it is important to invest in adequate infrastructure and resources and establish formal waste management systems that support the recovery and recycling of valuable materials. This can include establishing waste segregation and separation practices, building recycling facilities, and providing training and resources to support the formalization of the informal waste sector.

In addition to these efforts, waste-to-energy technologies, such as anaerobic digestion and gasification, offer opportunities to recover energy from waste materials. These technologies can be particularly relevant in areas with limited access to electricity and clean energy sources and can contribute to reducing greenhouse gas emissions and improving public health.

Promoting waste resource recovery in sub-Saharan Africa requires a multifaceted approach that considers the unique social, economic, and environmental contexts of different countries and communities. By working together to promote waste reduction, recycling, and energy recovery, it is possible to create more sustainable and resilient communities in sub-Saharan Africa.

2.3 ENVIRONMENTAL, ECONOMIC, AND SOCIAL SUSTAINABILITY

To minimize the environmental impacts and improve social well-being, environmental, social, and economic sustainability are the essential components to be incorporated into waste management systems.

These are the ways these three pillars can be integrated into waste management practices:

- *Environmental sustainability:* Waste management systems should be designed to minimize negative environmental impacts, such as greenhouse emissions, water, soil, and air pollution, and habitat destruction (Hajam et al., 2023). This can be achieved by promoting waste reduction, recycling, and energy recovery and by implementing proper waste storage, transport, and disposal practices that minimize environmental harm (Hajam et al., 2023). Additionally, waste management systems can incorporate green technologies and practices, such as renewable energy sources and sustainable materials, to further reduce environmental impacts (Kaygusuz, 2018).
- *Economic sustainability:* Waste management systems should be economically sustainable, meaning they should be financially viable over the long term (Avilés-Palacios & Rodríguez-Olalla, 2021). This can be achieved by promoting cost-effective waste management practices that balance the cost of collection, transport, and disposal with the benefits of waste reduction, recycling, and energy recovery (Wilson et al., 2006). Additionally, waste management systems can create economic opportunities through the creation of jobs, the generation of revenue from the sale of recyclable materials, and the development of new technologies and industries (Sharma et al., 2021b).
- *Social sustainability:* Waste management systems should also promote social sustainability by improving public health, reducing social inequalities, and promoting community engagement and empowerment (Roseland, 2000). This can be achieved by providing safe and sanitary waste management services that protect public health, promoting equitable access to waste management services, and involving communities in the design and implementation of waste management systems.

For Africa to achieve the SDGs, the three pillars of sustainability, as presented in Figure 2.4, should not be compromised.

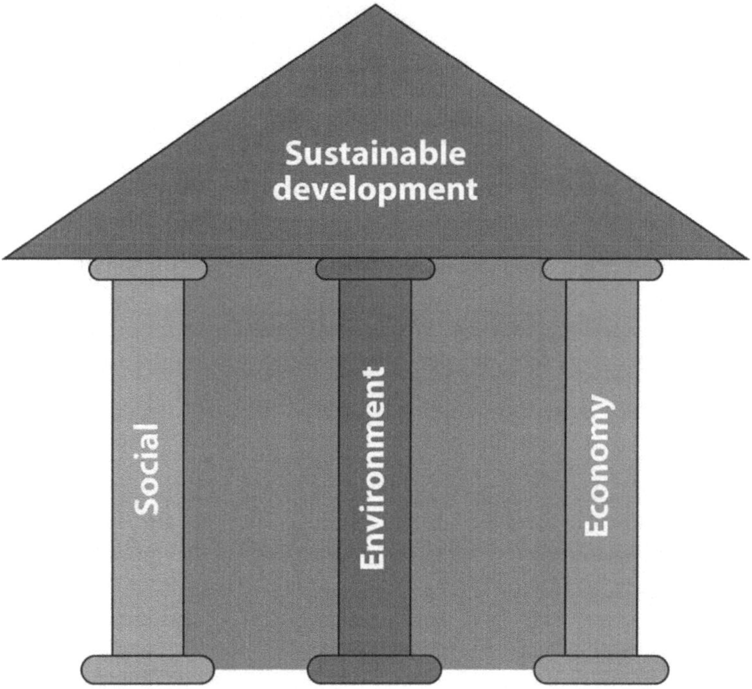

FIGURE 2.4 Pillars of sustainable development.

The waste management practices should uphold these pillars – social, environmental, and economic sustainabilities – and ensure the needs of the present are met without compromising the ability of future generations to meet their own needs. The necessity to preserve and safeguard ecosystems and natural resources is emphasized by the environmental pillar (Azam et al., 2023). The economic pillar focuses on promoting economic growth and development while ensuring resource efficiency and minimizing negative impacts on the environment (Soretz et al., 2023; Schoor et al., 2023). The social pillar emphasizes the significance of tackling social concerns, including social justice, poverty, and inequality, as well as making sure that individuals and their communities have a decent standard of living. All three of these interconnected and mutually reinforcing pillars must be considered if SDGs are to be met in Africa.

Sustainable development prioritizes the following features:

- Satisfying human needs
- Favouring a good quality of life through decent standards of living
- Sharing resources between rich and poor
- Acting with concern for future generations
- Looking at the "cradle-to-grave" impact when consuming
- Minimizing resource use, waste, and pollution

By integrating environmental, economic, and social sustainability principles into waste management practices, it is possible to create more sustainable and resilient communities that protect the environment, promote economic growth, and improve social well-being.

2.4 INTEGRATED SOLID WASTE MANAGEMENT

Integrated solid waste management (ISWM) is a comprehensive approach to managing solid waste that combines multiple waste management techniques, including waste reduction, recycling, composting, and energy recovery, in a coordinated and integrated manner (Memon, 2010). The goal of ISWM is to minimize the environmental, social, and economic impacts of waste while promoting sustainability.

ISWM involves several key components (Sharma et al., 2021b; Godfrey et al., 2019):

- Waste reduction: Waste reduction involves reducing the amount of waste generated through the use of efficient production processes, reducing packaging, and promoting the use of reusable and recyclable products.
- Recycling: Recycling involves the separation and recovery of materials that can be used again in the production of new products. This includes materials such as paper, plastics, metals, and glass.
- Composting: Composting involves the decomposition of organic waste to produce a nutrient-rich fertilizer that can be used in agriculture or landscaping.
- Energy recovery: Energy recovery involves the conversion of waste into energy through various methods, such as incineration or anaerobic digestion.
- Landfill management: Landfills are used as a last resort for waste that cannot be reduced, recycled, composted, or recovered for energy. ISWM includes proper landfill management practices, such as covering and compacting waste to reduce odours and methane emissions.

ISWM requires the cooperation and collaboration of various stakeholders, including waste generators, waste management companies, government agencies, and communities (Kim & Kim, 2023). Effective ISWM involves a balance between environmental protection, economic viability, and social equity, and must take into account the specific circumstances and needs of the community being served.

2.5 CIRCULAR ECONOMY FROM A PROCESS ENGINEERING POINT OF VIEW

Circular economy is a concept that emphasizes the importance of creating a closed-loop system for resources, where waste is minimized and materials are reused or recycled (Selomo, 2023; Cooper et al., 2024; Geissdoerfer et al., 2017). From a process engineering point of view, circular economy involves designing and optimizing industrial processes to minimize waste generation and maximize resource efficiency (Avraamidou et al., 2020).

When embarking on the implementation of the circular economy, process engineers can rely on the subsequent foundational principles, serving as a valuable framework for guidance. Designing for durability and recyclability is crucial in promoting sustainability within industrial processes. Products can be made to last longer and add less to landfill garbage by using materials that are simple to recycle and disassemble. Process intensification and waste heat recovery are two techniques that reduce waste and emissions from industrial operations, making them more efficient and ecologically beneficial. The adoption of closed-loop systems further enhances sustainability by reusing or recycling materials, reducing the need for new resources, and minimizing waste generation. Embracing renewable resources whenever possible helps to mitigate reliance on finite resources and promotes a more sustainable approach to production. The incorporation of eco-design principles into both process and product design results in minimizing environmental impacts and fosters a culture of sustainability throughout the industrial sector, ultimately contributing to a greener and more sustainable future.

Figure 2.5 presents the circular economy approach to move from a linear to a circular economy. Conducting a life cycle assessment can help identify the environmental impacts of a process or product and provide insights into areas where improvements can be made, while redesigning processes and products based on the results of the LCA can minimize waste and emissions, and incorporate closed-loop systems (Guillot, 2015). Integrating stakeholders like suppliers and customers can help encourage the adoption of circular economy principles throughout the supply

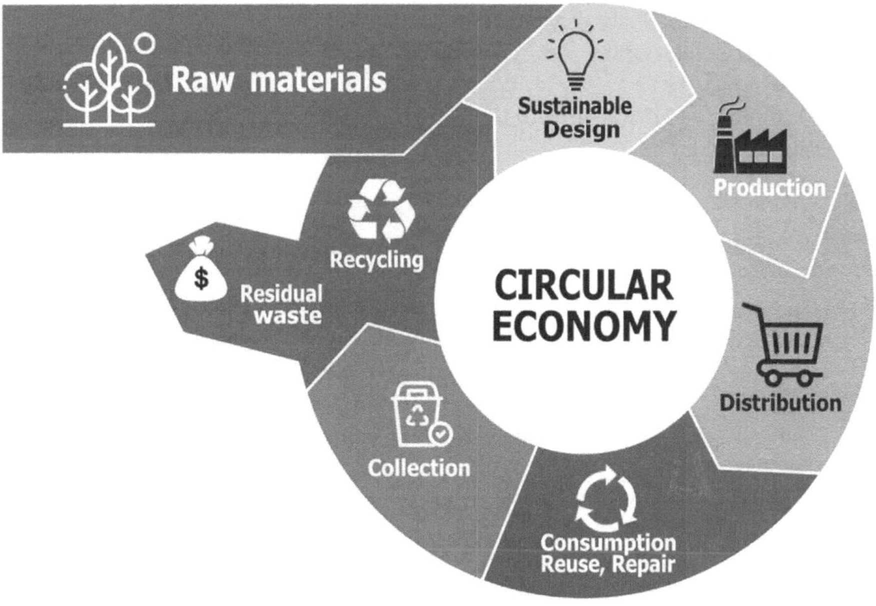

FIGURE 2.5 Circular economy approach.

Source: Guillot, 2015.

chain. New technologies, like process intensification and waste heat recovery, can optimize industrial processes and minimize waste.

The successful implementation of the circular economy requires a careful approach that considers the complete life cycle of products and processes, ranging from their beginning to their ultimate disposal (Guillot, 2015). The circular economy can improve the capability to augment sustainability and mitigate the detrimental impacts on the environment by optimizing industrial operations so that there is minimum waste production and maximum resource utilization (Guillot, 2015).

2.6 ZERO-WASTE MODEL

A zero-waste model is a holistic approach that aims to reshape contemporary modes of production and consumption by reducing waste and promoting active citizenship awareness (Guaran & Venturini, 2022). It recognizes waste as both a resource and a measure of inefficiency in society (Ko & Lu, 2023). The model seeks to emulate sustainable natural cycles, where all discarded materials are designed to become resources for others to use (Saribanon et al., 2022). Zero-waste principles can be applied in various contexts, such as environmental education, hospitality, fisheries, and microorganism growth (Ali & Shirazi, 2023; Heureux et al., 2022). The model encourages a systemic and circular approach, promoting a new perspective on sustainability and the redesign of our relationship with waste. It incorporates knowledge, attitude, and skill competencies and can help the circular economy grow. The zero-waste model can lower trash production, boost the value added to products, and aid in the creation of bioproducts that are renewable. Figure 2.6 illustrates a holistic zero waste model for a city (Zaman & Lehmann, 2011).

The key principles of the zero-waste model include:

- Designing and managing products and processes to eliminate waste volume and toxicity, and conserve and recover all resources (Guaran & Venturini, 2022).
- It is a circular system where the end of one product becomes the beginning of another, as opposed to a linear system where waste is generated from product consumption (Gaur et al., 2022).
- The model emphasizes resource recovery and reverse logistics to convert waste from one product into raw material for another (Heureux et al., 2022).
- Sustainable production through cradle-to-cradle design and product stewardship, collaborative and responsible consumption of natural resources, and zero-waste management through resource conservation (Yazdani & Lakzian, 2023).
- Promoting the use of scientific treatment techniques and mandatory refuse sorting and recycling (Zaman, 2017).

The zero-waste model encourages a systemic and circular approach to fundamentally reconsider and redesign our relationship with sustainability.

The zero-waste model is a holistic approach to waste management that prioritizes resource efficiency and sustainability (Zaman & Lehmann, 2011). By adopting this

FIGURE 2.6 A holistic zero-waste city model, with five interconnected key principles that need to be applied simultaneously.

Source: Zaman & Lehmann, 2011.

model, communities and businesses can reduce waste generation, promote circular economy principles, and minimize environmental impacts (Zaman & Lehmann, 2011). However, implementing a zero-waste model can be challenging and requires a coordinated effort from stakeholders across sectors.

2.7 5R OF SUSTAINABLE WASTE MANAGEMENT

The 5IRs of sustainable waste management are a set of principles that prioritize waste reduction and resource conservation. The principles of reduce, reuse, recycle, recover, and dispose are all included in the "5Rs" of environmentally friendly waste management. These guidelines seek to minimize waste production, maximize resource use, and lessen their negative effects on the environment. Reducing waste output overall may be achieved by eliminating waste at its source, such as by putting prevention and minimization techniques into practice. Product lifecycles are prolonged, and the requirement for new resources is decreased when resources are utilized repeatedly as opposed to being thrown away (Bhat et al., 2022). The process of extracting energy or valuable materials from waste through methods like waste-to-energy conversion is referred to as recovery (Kumar & Singh, 2023).

The hierarchy of 5Rs of sustainable waste management includes (Figure 2.7) (1) Reconsider/refuse, (2) Reduce, (3) Reuse, (4) Recycle, and (5) Repurpose (Kattiyapornpong et al., 2023). The concept of the hierarchy emphasizes waste avoidance before recycling and disposal (Singh & Hussain, 2021). It promotes reducing the use of raw materials as the most effective environmental approach to waste management. The hierarchy recommends prioritizing the reduction of raw material use, followed by reusing and recycling materials (Bui & Tseng, 2022). Incineration with heat recovery and landfilling are considered the last options in the hierarchy (Nilsen, 2020). The waste hierarchy has been introduced by the European Union and is aimed at turning member states into "recycling societies". It is a framework for sustainable resource management and encourages a circular economy approach.

The 5Rs of sustainable waste management as presented in Figure 2.8. aims to:

- *Refuse:* The first R is Refuse. This step involves saying no to items that are not necessary or that cannot be easily reused, recycled, or recovered. By refusing to use or purchase these items, the amount of waste generated can be reduced in the first place. This can be practiced by refusing to use single-use plastic bags, packaging, and straws.

FIGURE 2.7 5Rs of sustainable waste management.

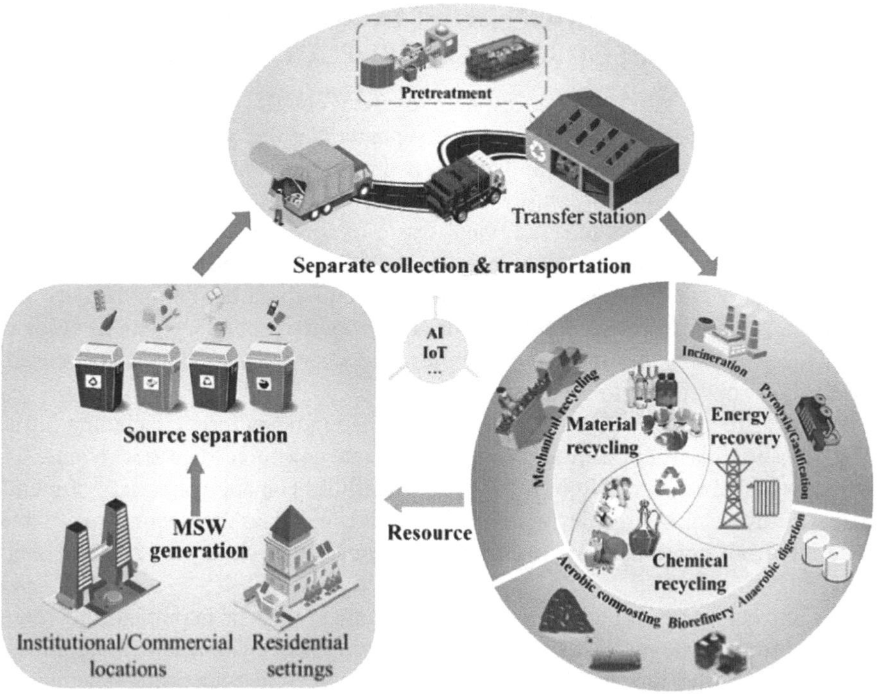

FIGURE 2.8 Valorization of solid waste to value-added products and resources.

Source: Zhang et al., 2021.

- *Reduce:* Reducing the amount of waste humans generate is the second and most important step in sustainable waste management. By consciously choosing to use less disposable products, utilizing products that have less packaging, and buying only what is required, waste can be reduced.
- *Reuse*: Instead of disposing of the unwanted items, there should be consideration to finding ways for reuse. This can be done by donating items and materials to others who can use them and repurposing items.
- *Recycle:* When reducing and reusing waste is not possible, the next step is to recycle. This step involves turning waste materials into new products. This is possible with wastes such as metal, plastic, and paper. Through recycling, the reduction of greenhouse gases and conservation of natural resources can be promoted and realized.
- *Repurpose/Recover:* This step involves recovering resources from waste that cannot be reduced, reused, or recycled. This step involves recovering resources from waste that cannot be reduced, reused, or recycled. This can be achieved through processes such as incineration, which generates energy from waste, or composting, which turns organic waste into a valuable soil amendment.

2.8 WASTE VALORIZATION (SHIFT FROM WASTE TREATMENT PLANTS TO RESOURCE RECOVERY PLANTS)

Waste valorization, also known as resource recovery, entails shifting from traditional waste treatment plants to resource recovery plants that recover valuable resources from waste. To create a more circular and sustainable economy, this shift is critical and will result in minimizing waste and maximizing the use of resources.

The common practice and traditional waste treatment plants dispose of waste, mainly hazardous wastes, either by burying it in landfills or burning it in incinerators. In practice, these methods reduce the volume of waste but do not recover any valuable resources from the waste streams. The resource recovery plants would normally focus on using innovative technologies to recover materials that can be reused, recycled, or converted into energy.

Resource recovery in plants can be implemented in many forms, including infrastructures for material recovery, composting, and anaerobic digestion plants. To extract valuable resources from wastes, these facilities can apply a variety of methods, such as turning organic waste into compost or biogas, separating recyclable materials from non-recyclable materials, and converting waste into energy (Babu et al., 2021).

By implementing resource recovery technologies, we can maximize the value obtained from waste materials and reduce the need for extraction of new resources, thereby mitigating the environmental impacts associated with mining activities (Bohra et al., 2022). Figure 2.8 shows a holistic approach to valorization of waste, focusing not only on recycling and reuse but resource recovery from the waste. In this approach, the waste is valorized and acts as a resource for the initial step of the process, closing the linear loop to a circular economy.

Shifting from waste treatment plants to resource recovery plants is a multifaceted process that encompasses various steps crucial for its success. Firstly, policy changes play a pivotal role in driving this transition. Governments can facilitate the shift by implementing policies and regulations geared towards waste reduction and resource recovery, such as imposing landfill taxes, advocating for extended producer responsibility, and setting ambitious recycling targets. Secondly, substantial investment in infrastructure is imperative.

Constructing resource recovery plants requires significant investment in new facilities and equipment, which can be undertaken by both governments and private entities to bolster this transformation. Moreover, public education plays a vital role in fostering acceptance and participation in resource recovery efforts. Public education and awareness about the benefits of resource recovery and proper ways and practices to sort and dispose of waste are essential components of this process. Additionally, collaboration among stakeholders is indispensable. Together, the public, commercial, and community sectors can open doors for resource recovery, the development of innovative technologies, and the construction of vital infrastructure. Lastly, innovation is key. Investing in research and development to pioneer new technologies and processes enhances the efficiency of resource recovery plants and contributes to the overall success of the transition. In essence, a comprehensive approach integrating policy changes, infrastructure investment, public education, collaboration among stakeholders, and innovation is crucial for effecting the shift from waste treatment to resource recovery plants.

2.9 POLLUTION PREVENTION AND CLEANER PRODUCTION CASE STUDIES WITHIN SUB-SAHARAN AFRICA

Pollution prevention and cleaner production are two interrelated concepts that aim to reduce the environmental impact of industrial activities and promote sustainable development (Prigozhin et al., 2023).

Pollution prevention refers to the systematic procedure of identifying, assessing, and implementing tactics aimed at diminishing or eradicating the production of waste and contaminants at their origin. The primary objective of pollution prevention is to avert ecological harm by decreasing both the volume and harmfulness of pollutants discharged into the surroundings. Pollution prevention methods encompass alterations in procedures, waste reduction, and the reconfiguration of products.

On the other side, planning and developing manufacturing methods and processes that are more effective, waste-free, and pollutant-free is known as cleaner production. The goal of cleaner production is to reduce waste and emissions while increasing resource utilization to lessen the damaging effects of industrial activity on the environment. Utilizing renewable energy sources, better process control, and cleaner technologies are some examples of cleaner production techniques.

Principles of sustainability, which highlight the need to balance economic, social, and environmental factors in all aspects of human activity, serve as the foundation for both pollution prevention and cleaner manufacturing (Prigozhin et al., 2023). By implementing cleaner production and pollution prevention strategies, industries and businesses can reduce their environmental impact and contribute to a more sustainable future (Agustina et al., 2023; Prigozhin et al., 2023).

Acknowledging the significance of pollution control and cleaner manufacturing, governments and international organizations have created regulations and initiatives to facilitate its execution. For example, the United Nations Industrial Development Organization (UNIDO) has established the Global Network for Resource Efficient and Cleaner Production (RECP), which aims to promote the adoption of cleaner production practices in developing countries (Penchel et al., 2023; Szilagyi & Mocan, 2018).

2.9.1 Evolution of Sustainable Production and Services

The evolution of sustainable production and services can be traced back to the emergence of the environmental movement in the 1960s and 1970s (Pak, 2011). This movement raised awareness about the negative environmental impacts of industrial activities and led to the development of new approaches to sustainable development (McMullin, 2023).

One of the first approaches to sustainable production was the concept of "eco-efficiency", which was introduced by the World Business Council for Sustainable Development in the 1990s (Muthu, 2020). This approach aimed to improve the environmental performance of businesses by reducing their use of natural resources and

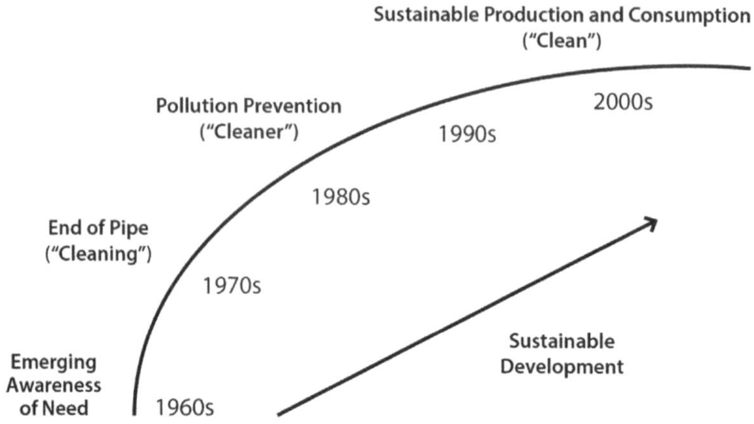

FIGURE 2.9 Evolution to sustainable production and services.

Source: Gavrilescu, 2004.

minimizing waste and pollution (Basuki, 2015). Figure 2.9 illustrates the evolution of sustainable processes and services (Gavrilescu, 2004).

Another important development in the evolution of sustainable production was the emergence of the "cradle-to-cradle" approach, which was developed by William McDonough and Michael Braungart in the early 2000s (Zhang et al., 2023; Mercader-Moyano & Porras-Pereira, 2023). This approach aimed to design products and services that were not only environmentally sustainable but also socially and economically beneficial (Fajarwati et al., 2023).

Sustainable services have evolved over time, with a growing emphasis on the social and environmental impacts of service delivery (Sarter, 2023). Numerous activities fall under the category of sustainable services, including sustainable travel, sustainable transportation, and sustainable food systems.

2.9.2 Integrated Approach to Cleaner Production

An integrated approach to cleaner production involves the application of environmental strategies at multiple levels, including macro, mezzo, and micro perspectives. This approach combines "Top-Down" measures such as financial subsidies, environmental efficiency improvement, and industrial restructuring with a "Bottom-Up" strategy of optimal technology selection. By integrating these measures, the targeted annual economic growth rate can be achieved without deterioration of the air environment, and air pollutant emissions can be significantly reduced. Upgrading air pollutant control technology is found to be more effective than other incentive measures, emphasizing the need for systematic optimization and technology choice shift. The integrated cleaner production strategies can achieve a strong decoupling effect, balancing economic development and environmental improvement in heavily air-polluted cities (Shao et al., 2021).

Cleaner production is a preventive environmental strategy that aims to make production processes more environmentally efficient by reducing resource and raw material use. It can help to reduce the environmental impact of different production systems and bring social, environmental, and economic benefits (Prigozhin et al., 2023). The continuous application of cleaner production involves integrating environmental strategies with processes, products, and services to increase efficiency and reduce risks to humans and the environment (de Oliveira et al., 2021). Cleaner production can be applied in various sectors, including educational facilities, where it can lead to significant cost savings, resource conservation, and environmental management (Nahui-Ortiz et al., 2020). To promote the adoption of cleaner production practices, a framework is proposed that includes actions in the government, scientific, and industrial sectors, such as national policies, legislation, tax incentives, clean technologies, and engagement of upper management (de Mello Santos et al., 2022).

A comprehensive approach to environmental management, the integrated approach to cleaner production (IACP) seeks to increase industrial process efficiency while reducing environmental impact. This method entails incorporating environmental factors into every stage of the production process, from product creation and design to trash disposal.

The IACP approach involves several key elements, as shown in Figure 2.10, including (Gavrilescu, 2004).

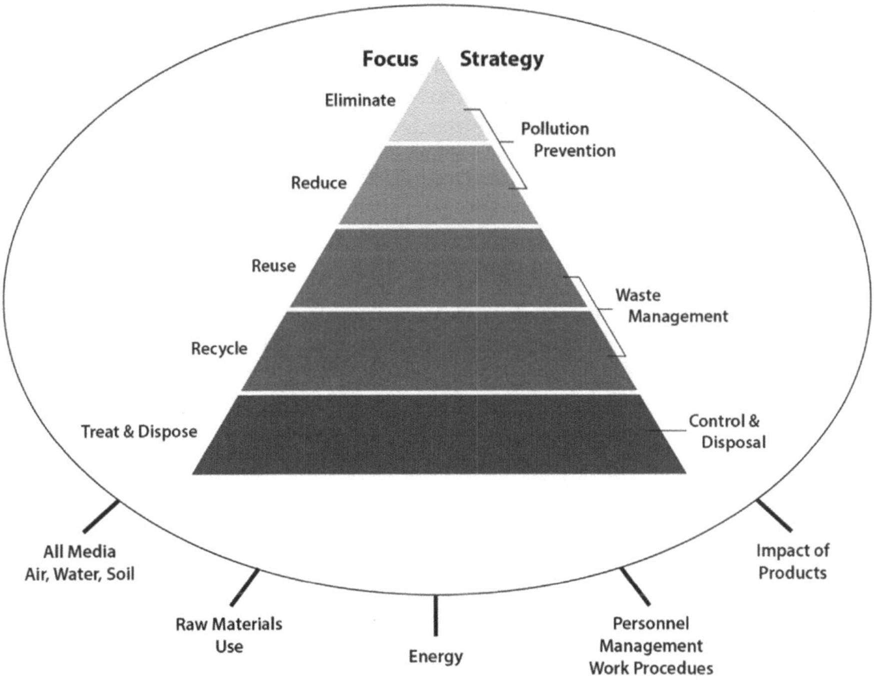

FIGURE 2.10 Integrated approach to Cleaner Production.

Source: Gavrilescu, 2004.

- *Prevention*
  The first step in the IACP approach is to prevent pollution and waste generation at the source by optimizing production processes, materials use, and product design.
- *Continuous improvement*
  The IACP approach is based on a continuous improvement cycle, where production processes are regularly reviewed and improved to minimize waste and pollution.
- *Stakeholder engagement*
  To guarantee that their opinions and concerns are taken into consideration, the strategy entails involving a variety of stakeholders in the environmental management process, such as communities, suppliers, consumers, and employees.
- *Life cycle thinking*
  The IACP approach takes a life cycle approach to environmental management, which considers the environmental impact of a product or service from its creation to its final disposal.
- *Resource efficiency*
  The IACP approach emphasizes the importance of resource efficiency, which involves minimizing the use of materials and energy while maximizing the output of products and services.

Businesses and organizations can enhance their environmental performance and reap financial rewards, including lower manufacturing costs, higher productivity, and enhanced reputation by implementing an IACP strategy (Agustina et al., 2023; Penchel et al., 2023). A wide range of industrial sectors, including manufacturing, construction, and agriculture, can use the IACP strategy. An effective instrument for encouraging sustainable development and raising companies' and organizations' environmental performance is the IACP strategy.

CASE STUDIES

Case Study 1

In sub-Saharan Africa, there are a few examples of the zero-waste model being implemented successfully. One of the most notable examples is the Zero Waste Academy in Accra, Ghana.

This academy is a community-based organization that provides training and education on waste management and promotes sustainable practices, such as composting and organic farming. The academy also provides waste collection services and operates a recycling facility, which processes plastics, glass, paper, and metal.

The Zero Waste Academy (ZWA) is a non-profit organization based in Accra, Ghana, that focuses on promoting sustainable waste management practices through education and community engagement. The organization was founded in 2018 by Mawuli Tse and Linda Akwaboah, both of whom are passionate about waste reduction and environmental conservation.

The ZWA aims to create a culture of zero waste in Ghana by providing education and training on sustainable waste management practices. The organization offers workshops, training programs, and community outreach events to educate individuals and organizations on how to reduce waste and minimize their impact on the environment. One of the key programs of the ZWA is the Zero Waste Ambassador Program, which trains individuals to become ambassadors for sustainable waste management practices in their communities. The program provides training on waste reduction, recycling, composting, and other sustainable waste management practices, and equips participants with the knowledge and tools to promote these practices in their communities.

The ZWA also partners with schools and universities to promote waste reduction and recycling among students. The organization provides educational resources and support to help schools establish recycling programs and implement sustainable waste management practices. In addition to education and training, the ZWA also engages in advocacy and awareness-raising activities to promote sustainable waste management practices in Ghana.

The organization works with government officials and policymakers to promote policies and regulations that support waste reduction and resource recovery. The ZWA has also launched a number of waste reduction initiatives in Accra, including the "Trashy Bags" project, which transforms plastic waste into fashionable bags and accessories, and the "Green Box" project, which provides households and businesses with composting bins to help them turn organic waste into nutrient-rich compost. Overall, the Zero Waste Academy is an important organization in Ghana's efforts to promote sustainable waste management practices and reduce the environmental impact of waste. Through its education and outreach programs, the ZWA is helping to create a culture of zero waste in Ghana and inspire individuals and organizations to take action to protect the environment.

Case Study 2

Another example is the Zero Waste Kigali initiative in Rwanda. This initiative aims to create a zero-waste city by promoting sustainable practices such as composting and recycling, and by working with businesses and industries to reduce waste at the source. The initiative has been successful in reducing waste in Kigali, with the city now producing less than 10% of its waste in landfills.

The Zero Waste Kigali Initiative is a community-based organization in Rwanda that aims to promote sustainable waste management practices and reduce waste in the community of Kigali. The organization was founded in 2015 by Jean Paul Ndagijimana, a local resident who was concerned about the environmental impact of waste in his community.

The initiative is based on the principles of the zero-waste movement, which advocates for a circular economy where waste is minimized and resources are conserved through strategies such as reducing, reusing, recycling, and composting.

The Zero Waste Kigali Initiative has implemented several programs and initiatives to promote sustainable waste management practices in the community. These include:

- Waste segregation: The initiative has established a waste segregation program, where residents are encouraged to separate their waste into different categories, such as organic waste, plastics, and paper. The initiative provides education and training to help residents understand the importance of waste segregation and how to properly sort their waste.
- Composting: The initiative has established a composting program, where organic waste is collected and composted to produce nutrient-rich soil that can be used for agriculture. The initiative provides training and support to residents to help them set up composting systems in their homes.
- Recycling: The initiative has established a recycling program, where plastic waste is collected and sent to a recycling facility to be processed into new products. The initiative has also established a partnership with a local artist who uses recycled materials to create art pieces.
- Community outreach: The initiative conducts community outreach activities to promote awareness about sustainable waste management practices. These activities include workshops, seminars, and community clean-up events.

The Zero Waste Kigali Initiative has achieved significant success in promoting sustainable waste management practices in the community. Through its programs and initiatives, the initiative has reduced the amount of waste generated in the community, improved the health and well-being of residents, and created economic opportunities through the recycling and composting programs. Overall, the Zero Waste Kigali Initiative is an inspiring example of how communities can come together to promote sustainable waste management practices and reduce their environmental impact.

Case Study 3

In South Africa, the zero-waste model is being implemented in a number of communities through initiatives such as the Wild Coast Zero Waste Project. This project promotes composting and recycling and provides education and training on waste reduction and sustainable living. The project has been successful in reducing waste and improving the local environment and has become a model for other communities in the region.

The Wild Coast Zero Waste Project is a non-profit organization based in the Eastern Cape Province of South Africa, which aims to promote sustainable waste management practices and reduce waste in the region. The organization was founded in 2011 by local residents who were concerned about the environmental impact of waste in their communities. The Wild Coast Zero Waste Project focuses on promoting the principles of the zero-waste movement, which advocates for a circular economy where waste is minimized and resources are conserved through strategies such as reducing, reusing, recycling, and composting.

The organization has implemented a number of programs and initiatives to promote sustainable waste management practices in the region. These include:

- Waste reduction and segregation: The organization promotes waste reduction and segregation through education and awareness-raising activities. The organization provides training and support to community members to help them reduce their waste and properly sort it into different categories.
- Composting: The organization promotes composting as a way to turn organic waste into nutrient-rich soil that can be used for gardening and agriculture. The organization provides training and support to community members to help them set up composting systems in their homes.
- Recycling: The organization promotes recycling as a way to reduce waste and conserve resources. The organization has established partnerships with local recycling companies to collect and process recyclable materials.
- Education and outreach: The organization conducts education and outreach activities to promote awareness about sustainable waste management practices. These activities include workshops, seminars, and community clean-up events.
- Advocacy: The organization advocates for policies and regulations that support sustainable waste management practices at the local and national levels.

The Wild Coast Zero Waste Project has achieved significant success in promoting sustainable waste management practices in the region. Through its programs and initiatives, the organization has reduced the amount of waste generated in the region, improved the health and well-being of residents, and created economic opportunities through the recycling and composting programs. By implementing strategies such as waste reduction, composting, and recycling, the organization is helping to create a more sustainable future for the region and inspiring others to take action to protect the environment.

2.10 PROBLEMS

1. Define circular economy. [2]
2. Explain the 3R concept of waste management. (Use a diagram to support your answer) [4]
3. "The earth is losing its carrying capacity due to unsustainable production and consumption" – Can engineers help regain the carrying capacity of the earth? Do you agree or disagree? State the reasons for your answer. [6]
4. Figure 2.11 shows the process to produce concentrated sulphuric acid. The first step of the process produces sulphur dioxide, which is used as the raw material for sulphuric acid production. Sulphur is melted and burned at high temperatures to produce sulphur dioxide.

Sulphur dioxide is then reacted with oxygen in the presence of a catalyst (vanadium pentoxide, v_2O_5) to produce sulphur trioxide. The sulphur trioxide gas is then absorbed into concentrated sulphuric acid to form oleum, which is a mixture of sulphuric acid and sulphur trioxide, carried out through a series of absorption towers. The oleum is then diluted with water to produce concentrated sulphuric acid at the desired strength.

The process generates the following by-products and pollutants:

- SO_2 as a by-product
- Nitrogen oxides (NOx) during the combustion process
- Particulate matter, which are fine particles released into the atmosphere.
- Hazardous waste, such as spent acid, sludge, and ash.

4.1. Discuss how you would implement a circular economy approach for this sulphuric acid manufacturing plant. Use diagram/s to support your answer. [15]

5. Why is Sustainable Engineering important? [3]
6. How can sustainable engineering be applied in engineering field? Provide an example.

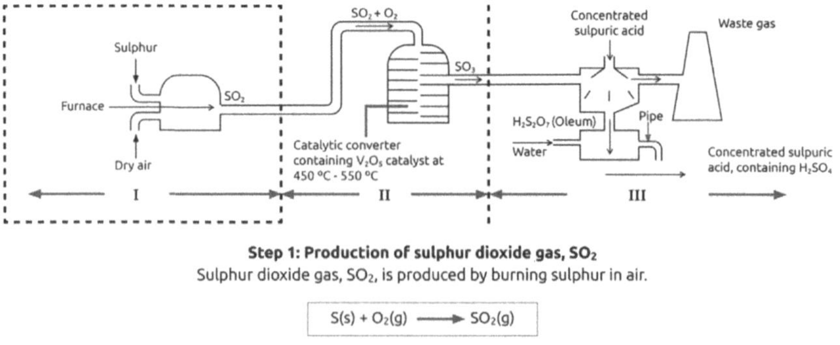

FIGURE 2.11 Production of concentrated sulphuric acid.

[6]

7. What are the three pillars of sustainable development and how do they interact with each other? [8]
8. Sustainable Development Goal 13 is to limit and adapt to climate change. It is one of the 17 SDGs established by the United Nations General Assembly in 2015 (Figure 2.12). The cement industry is responsible for about 8% of planet-warming carbon dioxide emissions – far more than global carbon

FIGURE 2.12 United Nations Sustainable Development Goals.

FIGURE 2.13 Pile of rubble.

emissions from aviation. If the cement industry were a country, it would be the third-largest emitter of carbon dioxide in the world.

8.1. As an Engineer, how would you propose a solution to address climate change related to the industry? [8]

9. The construction sector is one of the sectors that impact the environment significantly. A company has been contracted to demolish a big building and then build a school in a city.

One of the wastes generated in the process of demolishing is solid waste in the form of a rubble, and steel material as shown in Figure 2.13.

9.1. Discuss the benefits of circular economy. [5]

9.2. How will you apply circular economy to manage the wastes generated? (Use a diagram to support your answer) [6].

REFERENCES

Adedara, M. L., Taiwo, R., & Bork, H. R. (2023, April). Municipal solid waste collection and coverage rates in sub-Saharan African countries: A comprehensive systematic review and meta-analysis. In *Waste* (Vol. 1, No. 2, pp. 389–413). MDPI.

Adeniran, A. A., Ayesu-Koranteng, E., & Shakantu, W. (2022). A review of the literature on the environmental and health impact of plastic waste pollutants in sub-Saharan Africa. *Pollutants*, *2*(4), 531–545.

Agbabiaka, H. I., Musa, N. S. A., & Olugbamila, O. B. (2022). Creating a recycled based society through solid waste transformation: A step towards achieving sustainable development goals (SDGS) 1 and 11 in sub-Saharan African. *The Journal of Solid Waste Technology and Management*, *48*(3), 415–426.

Agustina, D., Wicaksono, A. D., & Meidiana, C. (2023). The influencing factors in cleaner production adoption on the aluminium processing industry. *JEMIS (Journal of Engineering & Management in Industrial System)*, *11*(1), 14–25.

Akinjogbin, I. O., Sridhar, M. K. C., & Coker, A. O. (2020). Design and fabrication of a material recovery facility for institutional wastes to mitigate carbon emissions in South-West Nigeria. *International Journal of Engineering Applied Sciences and Technology*, *5*(4), 32–38. http://doi.org/10.33564/IJEAST.2020.v05i04.005

Akpan, V. E., & Olukanni, D. O. (2020). Hazardous waste management: An African overview. *Recycling*, *5*(3), 15.

Ali, S., & Shirazi, F. (2023). The paradigm of circular economy and an effective electronic waste management. *Sustainability*, *15*(3), 1998.

Amankwa, M. O., Tetteh, E. K., Mohale, G. T., Dagba, G., & Opoku, P. (2021). The production of valuable products and fuel from plastic waste in Africa. *Discover Sustainability*, *2*, 1–11.

Apazhev, A. K., Shekikhachev, Y. A., Fiapshev, A. G., Shekikhacheva, L. Z., & Fiapshev, B. A. (2022, December). Environmentally oriented disposal of waste from agricultural enterprises in a biomethane plant. In *IOP conference series: Earth and environmental science* (Vol. 1112, No. 1, p. 012023). IOP Publishing.

Avilés-Palacios, C., & Rodríguez-Olalla, A. (2021). The sustainability of waste management models in circular economies. *Sustainability*, *13*(13), 7105.

Avraamidou, S., Baratsas, S. G., Tian, Y., & Pistikopoulos, E. N. (2020). Circular economy: A challenge and an opportunity for process systems engineering. *Computers & Chemical Engineering*, *133*, 106629.

Azam, M., Boubaker, S., & Hunjra, A. I. (2023). Dynamic institutions as pillars for sustainable development. In *Governance in the developing world: Transnational insights on economic development* (pp. 1–22). Springer Nature.

Babu, R., Veramendi, P. M. P., & Rene, E. R. (2021). Strategies for resource recovery from the organic fraction of municipal solid waste. *Case Studies in Chemical and Environmental Engineering, 3*, 100098.

Basuki, B. (2015). Eco-efficiency and sustainable development as efforts to produce environmentally friendly product: An exploratory case study. *Issues in Social & Environmental Accounting, 9*(3).

Bhat, M. A., Adil, A. W., Sikander, B. M., Lone, Y., & Malik, J. A. (2022). Waste management technology for sustainable agriculture: Waste management. In *Research anthology on strategies for achieving agricultural sustainability* (pp. 379–398). IGI Global.

Bohra, V., Ahamad, K. U., Kela, A., Vaghela, G., Sharma, A., & Deka, B. J. (2022). Energy and resources recovery from wastewater treatment systems. In *Clean energy and resource recovery* (pp. 17–36). Elsevier.

Buda, G. (2022). Seven businesses using principles of circular economy in sub-Saharan Africa: Results of field research in Uganda. *The Hungarian Journal of African Studies, 16*, 5.

Buda, G., & Ricz, J. (2023). Industrial symbiosis and industrial policy for sustainable development in Uganda. *Review of Evolutionary Political Economy*, 1–25.

Bui, T. D., & Tseng, M. L. (2022). Understanding the barriers to sustainable solid waste management in society 5.0 under uncertainties: A novelty of socials and technical perspectives on performance driving. *Environmental Science and Pollution Research, 29*(11), 16265–16293.

Cooper, A. K., Kruger, T., Godfrey, L. K., & Napier, M. (2024). *The circular economy, human settlements and municipal government*. Alive2Green.

Ddiba, D. (2020). *Exploring the circular economy of urban organic waste in sub-Saharan Africa: Opportunities and challenges* [Doctoral dissertation, KTH Royal Institute of Technology].

Debrah, J. K., Teye, G. K., & Dinis, M. A. P. (2022). Barriers and challenges to waste management hindering the circular economy in sub-Saharan Africa. *Urban Science, 6*(3), 57.

de Mello Santos, V. H., Campos, T. L. R., Espuny, M., & de Oliveira, O. J. (2022). Towards a green industry through cleaner production development. *Environmental Science and Pollution Research*, 1–22.

de Oliveira, J. A., da Cruz, A. J. G., Longati, A. A., & Fidanza, L. B. (2021). Cleaner production (CP). *Life Cycle Engineering and Management of Products: Theory and Practice*, 13–44.

Duodu, E., & Mpuure, D. M. N. (2023). International trade and environmental pollution in sub-Saharan Africa: Do exports and imports matter? *Environmental Science and Pollution Research, 30*(18), 53204–53220.

Fajarwati, A. A. S., Wulandari, A. A. A., Caroline, O. S., Ruki, U. A., Fanthi, R., & Fathoni, A. F. C. A. (2023). Sustainable design strategy through upcycling waste wood material and creating a digital marketplace for micro crafters community. In *IOP conference series: Earth and environmental science* (Vol. 1169, No. 1, p. 012037). IOP Publishing.

Faried, A. I., Sembiring, R., Wardhani, N. S., & Hasanah, U. (2023). Transition to a circular economy-inspired waste management system. *World Journal of Advanced Research and Reviews (WJARR), 19*(1), 145–153.

Gaur, A., Gurjar, S. K., & Chaudhary, S. (2022). Circular system of resource recovery and reverse logistics approach: Key to zero waste and zero landfill. In *Advanced organic waste management* (pp. 365–381). Elsevier.

Gavrilescu, M. (2004). Cleaner production as a tool for sustainable development. *Environmental Engineering & Management Journal (EEMJ), 3*(1).

Geissdoerfer, M., Savaget, P., Bocken, N. M., & Hultink, E. J. (2017). The circular economy: A new sustainability paradigm? *Journal of Cleaner Production, 143*, 757–768.

Geme, T., Nijman, E., Ntawuhiganayo, E. B., & Negesa, D. (2023). Stakeholder knowledge and perceptions of the circular economy in Ugandan cities. *Frontiers in Sustainability*, *4*, 1117814.

Godfrey, L., Ahmed, M. T., Gebremedhin, K. G., Katima, J. H., Oelofse, S., Osibanjo, O., Richter, U. H., & Yonli, A. H. (2019). Solid waste management in Africa: Governance failure or development opportunity. *Regional Development in Africa, 235.*

Guaran, A., & Venturini, F. (2022). Zero Waste for geographical education on sustainability. *Geography Notebooks*, *5*(1), 73–85.

Guillot, J. D. (2015). *Circular economy: Definition, importance and benefits*. European Parliament. PDF-dokumentti. Saatavissa. https://www.europarl.europa

Gul, S., Ullah, Q., & Qasim, M. (2022). Bio-waste management legislation regulations and policies: A case study of Pakistan. *Pakistan Journal of International Affairs*, *5*(3).

Guman, O., & Wegner-Kozlova, E. (2020). *Waste management based on circular economy principles*. EDP Sciences. https://doi.org/10.1051/e3sconf/202017704014

Hajam, Y. A., Kumar, R., & Kumar, A. (2023). Environmental waste management strategies and vermi transformation for sustainable development. *Environmental Challenges*, 100747.

Heureux, A. M. C., Matsumoto, T. K., & Keith, L. M. (2022). Toward a zero-waste model: Potential for microorganism growth on agricultural waste products in Hawaii. *Algal Research*, *62*, 102640.

Ibrahima, S. (2019). *Wastes management in African cities*. Springer. https://doi.org/10.1007/978-981-13-3471-9_6

Idowu, I. A., Atherton, W., Hashim, K., Kot, P., Alkhaddar, R., Alo, B. I., & Shaw, A. (2019). An analyses of the status of landfill classification systems in developing countries: Sub Saharan Africa landfill experiences. *Waste Management*, *87*, 761–771.

Kattiyapornpong, U., Ditta-Apichai, M., & Chuntamara, C. (2023). Sustainable food waste management practices: Perspectives from five-star hotels in Thailand. *Sustainability*, *15*(13), 10213.

Kaygusuz, K. (2018). Green chemistry and green energy technologies for environmental friendly sustainable development. *Journal of Engineering Research and Applied Science*, *7*(2), 953–964.

Kim, N. I., & Kim, K. C. (2023). A framework to develop and improve construction and demolition waste management through the collaborative action of organizations. *Journal of Energy, Environmental & Chemical Engineering*, *8*(2), 26–39.

Ko, W. H., & Lu, M. Y. (2023). A model to establish a zero food waste competence scale for hospitality students. *Journal of Hospitality, Leisure, Sport & Tourism Education*, *32*, 100422.

Kumar, M., & Singh, S. K. (2023). *Futuristic projects in energy and automation sectors: A brief review of new technologies driving sustainable development* (pp. 21, 107–127). Bentham Science Publishers.

Li, C. H., Lee, T. T., & Lau, S. S. Y. (2023). Enhancement of municipal solid waste management in Hong Kong through innovative solutions: A review. *Sustainability*, *15*(4), 3310.

Laovisutthichai, V., Lu, W. and Bao, Z. (2022). Design for construction waste minimization: guidelines and practice. *Architectural Engineering and Design Management*, *18*(3), 279–298.

McMullin, C. (2023). Expectations versus reality: The sustainability of co-production approaches over time. *Public Management Review*, 1–20.

Memon, M. A. (2010). Integrated solid waste management based on the 3R approach. *Journal of Material Cycles and Waste Management*, *12*, 30–40.

Mercader-Moyano, P., & Porras-Pereira, P. (Eds.). (2023). Definition of the sustainability criteria in the materials used: Cradle-to-cradle certified products. In *Circular economy in emergency housing: Eco-efficient prototype design for Subaşi refugee camp in Turkey and Maicao refugee camp in Colombia: A research strategy of climate change* (pp. 67–71). Springer Nature Switzerland.

Mombo, F., & Bigirwa, D. (2017). The role of sub-Saharan Africa countries' households waste charges on sustainable cities development. *International Journal of Waste Resources, 7*(265), 2.

Moyen Massa, G., & Archodoulaki, V. M. (2023). Electrical and electronic waste management problems in Africa: Deficits and solution approach. *Environments, 10*(3), 44.

Muthu, S. S. (2020). *Assessing the environmental impact of textiles and the clothing supply chain*. Woodhead Publishing.

Mwanza, B. G., Mbohwa, C., & Telukdarie, A. (2018). The influence of waste collection systems on resource recovery: A review. *Procedia Manufacturing, 21*, 846–853.

Nahui-Ortiz, J., Camarena-Gamarra, C., & Mendoza, A. (2020). *Integrated application of cleaner production on a university campus*. LACCEI Inc.

Ndam, S., Touikoue, A. F., Chenal, J., Baraka Munyaka, J. C., Kemajou, A., & Kouomoun, A. (2023, July). Urban governance of household waste and sustainable development in sub-Saharan Africa: A study from Yaoundé (Cameroon). In *Waste* (Vol. 1, No. 3, pp. 612–630). MDPI.

Nilsen, H. R. (2020). The hierarchy of resource use for a sustainable circular economy. *International Journal of Social Economics, 47*(1), 27–40.

Njenga, M., & Mendum, R. (2018). *Recovering bioenergy in Sub-Saharan Africa: Gender dimensions, lessons and challenges*. International Water Management Institute (IWMI). CGIAR Research Program on Water, Land and Ecosystems (WLE).

Ogutu, F. A., & Kathambi, B. (2022). Recycling gap, Africa's perspective for sustainable waste management. In *Recycling strategy and challenges associated with waste management towards sustaining the world*. IntechOpen.

Ossei-Bremang, R. N., Adjei, E. A., Kemausour, F., & Gyenin, E. K. (2023). *Industrial symbiosis and circularization: Optimal shelf life of waste-based briquettes*. https://doi.org/10.21203/rs.3.rs-2981891/v1

Pak, M. S. (2011). Environmentalism then and now: From fears to opportunities, 1970–2010. *Environmental Science & Technology, 54*, 5–9.

Penchel, R. A., Aldaya, I., Marim, L., dos Santos, M. P., Cardozo-Filho, L., Jegatheesan, V., & de Oliveira, J. A. (2023). Analysis of cleaner production performance in manufacturing companies employing artificial neural networks. *Applied Sciences, 13*(6), 4029.

Prigozhin, V., Zheng, W., & Lee, F. (2023). Implementing cleaner production technologies: General aspects. In *E3S web of conferences* (Vol. 392, p. 01006). EDP Sciences.

Rayfuse, R. (2016). Principles of international environmental law applicable to waste management. In *Waste management and the green economy* (pp. 11–32). Edward Elgar Publishing.

Roseland, M. (2000). Sustainable community development: Integrating environmental, economic, and social objectives. *Progress in Planning, 54*(2), 73–132.

Saribanon, N., Akman, A., Rijkie, K. M., Masulili, A. S., Ginting, R., Ananta, B., & Siregar, M. A. P. (2022). Developing a small enterprise business model with zero waste management systems approach. *Journal of Social Political Sciences, 3*(3), 281–290.

Sarter, E. K. (2023). Public services and the challenge of sustainability. In *Understanding public services* (pp. 103–118). Policy Press.

Schoor, M., Arenas-Salazar, A. P., Torres-Pacheco, I., Guevara-González, R. G., & Rico-García, E. (2023). A review of sustainable pillars and their fulfillment in agriculture, aquaculture, and aquaponic production. *Sustainability, 15*(9), 7638.

Selomo, T. G. (2023). *Solid waste management and selection of a solid waste disposal site in the Mankweng cluster* [Doctoral dissertation, Polokwane Local Municipality].

Shao, J., Xiang, N., Zhang, Y., Li, X., & Liang, G. (2021). Dynamic simulation of integrated cleaner production strategies towards high quality development in a heavily air-polluted city in China. *Sustainability*, *13*(16), 8951.

Sharma, H. R., Bhardwaj, B., Sharma, B., & Kaushik, C. P. (2021b). Sustainable solid waste management in India: Practices, challenges and the way forward. *Climate Resilience and Environmental Sustainability Approaches: Global Lessons and Local Challenges*, 319–349.

Singh, S., & Hussain, C. M. (2021). Zero waste hierarchy for sustainable development. In *Concepts of advanced zero waste tools* (pp. 123–142). Elsevier.

Soretz, S., Nodehi, M., & Taghvaee, V. M. (2023). *Sustainability and sustainable development concepts: Defining the perspectives and pillars*. Reference Module in Social Sciences.

Szilagyi, A., & Mocan, M. (2018). Scaling up resource efficiency and cleaner production for an sustainable industrial development. *Procedia-Social and Behavioral Sciences*, *238*, 466–474.

Wikurendra, E. A., Syaifuddin, A., Herdiani, N., & Nurika, G. (2023). Forecast of waste generated and waste fleet using linear regression model. *Polish Journal of Environmental Studies*, *32*(2), 1867–1876.

Wilson, D. C., Velis, C., & Cheeseman, C. (2006). Role of informal sector recycling in waste management in developing countries. *Habitat International*, *30*(4), 797–808.

Yazdani, S., & Lakzian, E. (2023). Conservation; waste reduction/zero waste. In *Pragmatic engineering and lifestyle: Responsible engineering for a sustainable future* (pp. 131–152). Emerald Publishing Limited.

Zaman, A. U. (2017). A strategic framework for working toward zero waste societies based on perceptions surveys. *Recycling*, *2*(1), 1.

Zaman, A. U., & Lehmann, S. (2011). Challenges and opportunities in transforming a city into a "zero waste city". *Challenges*, *2*(4), 73–93.

Zhang, X., Liu, C., Chen, Y., Zheng, G., & Chen, Y. (2021). Source separation, transportation, pretreatment, and valorization of municipal solid waste: A critical review. *Environment, Development and Sustainability*, 1–43.

Zhang, X., Liu, X., Zhang, Y., Xu, X., Xiao, J., & Luh, D. B. (2023). Assessing the feasibility of practical cradle to cradle in sustainable conceptual product design. *Sustainability*, *15*(8), 6755.

3 Fundamentals of Waste Management and Their Application

Thabo Falayi and Freeman Ntuli

3.1 INTRODUCTION

One of the key factors in waste management is determining the quantity of waste and its characteristics. This is essential in order to implement the 4Rs (Reduce, Reuse, Recycle, and Recover) as a means of driving towards a zero-waste circular economy. The effectiveness of technologies for processing waste is dependent on its composition and physical properties, and this is a key factor in determining the most appropriate approach to waste valorization. The chapter introduces the general waste handling methods and their applicability.

3.2 WASTE QUANTIFICATION AND SAMPLING

The sampling of waste is an important step in the quantification of waste. Various sampling methods have been reported. Random sampling over a period of time has been reported in literature for municipal solid waste (Ramakrishna et al., 2016). This allows to account for variations within the set time. The stratified random sampling of areas based on income has also been used for sampling purposes of household waste (Nakin et al., 2022). Stratification can be based on residential structure, population income, level of literacy, and type of collection, only to mention a few (Weichgrebe et al., 2017). Other sampling methods also include simple random sampling, systematic sampling, ranked set sampling, and sequential sampling (Weichgrebe et al., 2017). It is important to note that there is no consistency in sampling methods and frequency in the reported literature; this therefore talks about adaptability of the accepted methods (Nell et al., 2022). Nell et al. (2022) reported that various authors had varying sampling ratios ranging from 1 bag per 64 – 563 households. However, the standardized international method for sampling of unprocessed solid waste is provided by the ASTM D5231-92 (2003). The sampling is therefore supposed to be dependent on statistical criterion, and sampling should be random and over a period of time.

There are various quantification methods for waste, and these are industry and country specific. Waste quantification is important for planning and is usually an

DOI: 10.1201/9781003377856-3

indicator of country or industry best practice (Hassan et al., 2018). In the construction, the quantification methods can broadly be categorized into soft and hard methods. The soft methods include interviews with construction personnel and statistics-based estimates. The hard methods include Material Flow Analysis Approach (MFA) and Sorted and Weighed the Waste Material (SWWM). The MFA tracks the quantity of materials and their composition coming into a construction site and then estimates when the material will come out as waste (Hassan et al., 2018). The most accurate method for quantification of waste is the SWWM. The method uses a sampling method to get a representative of the waste that is exiting a site. The only challenge with the method is that it is cumbersome. The statistics-based method is dependent on the availability of data within the site or country. Case studies have also been utilized to quantify the waste that is generated through field observations, sorting and weighing on-site, truckload records, and surveys (Villoria-Sáez et al., 2020). There are a number of artificial intelligence tools that have been developed to estimate waste in the construction industry, and these include net waste tool, DeconRCM, and Cype software, among others (Villoria-Sáez et al., 2020). Another method used in the construction industry is the material lifetime analysis. The method uses the current production quantities of key construction materials and their shelf life to estimate quantities (Berge & von Blottnitz, 2022). Municipal solid waste is based on collection of data from selected household's and then extrapolating the data using census data of the number of households (Nakin et al., 2022). The bulk of the reported data on household waste is from research work and not from the municipal authorities although some municipalities quantify waste using number of loads and number of bags (Crosset, 2014). These can then be used to calculate the mass of the waste given the tare weights of waste collectors and bulk density of the waste.

3.3 WASTE CHARACTERIZATION

The characterization of solid waste is an expense and is therefore dependent on what the waste is to be used for. There are many tests that are used for the characterization, some of which are described in this section. Waste characterization is an important determinant of waste management systems.

3.3.1 Bulk Density

Bulk density is a good indicator of the ease with which waste can be compacted. This is useful for the quantity that can be carried by waste refuse trucks at any given time or the amount of space required to landfill the waste.

3.3.2 Proximate Analysis

Proximate analysis indicates the moisture content, volatile matter, fixed carbon, and ash the waste contains. There are several methods that can be used for this characterization, and these include the American Society for Testing and Materials (ASTM) methods E 872, E 1756, and D 1102 for volatile matter, moisture content, and ash, respectively.

Moisture is defined as the weight of water contained in the waste. Moisture affects a number of processes that the waste is subjected to, making it a critical parameter for assessment. In thermal waste treatment, elevated moisture content results in increased energy costs for moisture removal. Moisture content is calculated as follows:

$$\text{Moisture content} = \frac{\text{wet weight} - \text{dry weight}}{\text{weight weight}} \tag{3.1}$$

giving a moisture content on a wet basis, while the dry basis is calculated as

$$\text{Moisture content} = \frac{\text{wet weight} - \text{dry weight}}{\text{dry weight}} \tag{3.2}$$

The drying is usually done at 75°C or 105°C till a constant weight is obtained.

Volatile matter is the solid capable of combustion which evaporates at temperatures around 950° C. Fixed carbon is the combustible portion of waste after the removal moisture, volatile matter, and ash content. It is calculated as follows

$$\text{Fixed Carbon\%} = 100\% - \%\text{moisture} - \%\text{ash} - \%\text{Volatile matter} \tag{3.3}$$

3.3.3 Ultimate Analysis

This is used for elemental analysis and a number of instruments can be used for such purposes, including carbon and sulphur analyzers and X-ray fluorescence spectroscopy. The ultimate analysis can be used to calculate the heat value of waste using the extended Dulong equation

$$\frac{\text{Btu}}{\text{lb}} = 144\text{C} + 672\text{H} + 6.2\text{O} + 41.4\text{S} - 10.8\text{N} \tag{3.4}$$

where C, H, O, S, and N are again the weight percentages (dry basis) of carbon, hydrogen, oxygen, sulphur, and nitrogen, respectively, in the combustible fraction of the fuel (Worrell & Vesilind, 2010). The conversion of Btu to Joules is given by multiplying the Btu figure by 1054.

Example 3.1

The waste from a municipal has the composition as shown in Table 3.1.

Determine the empirical formula of the waste on a dry and the moisture content of the waste using the information given in Table 3.2:

TABLE 3.1
Composition of Municipal Waste

Component	Mass %	% Moisture Content
Newspaper	15.00	6
Other papers	24.00	5
Cardboard	33.00	5
Glass	4.20	0.5
Plastics	0.49	2
Yard wastes	17.97	60
Food wastes	1.67	60
Dirt	2.01	8

TABLE 3.2
Ultimate Analysis of Waste

	% by mass (dry basis)					
Ultimate analysis	**C**	**H**	**O**	**N**	**S**	**Ash**
Yard wastes	48	6	38	3.0	0.3	4.7
Wood	50	6	43	0.2	0.1	0.7
Food wastes	50	6	38	3.0	0.4	2.6
Paper	44	6	44	0.3	0.2	5.5
Cardboard	44	6	44	0.3	0.2	5.5
Plastic	60	7	23			10.0
Textiles	56	7	30	5.0	0.2	1.8
Rubber	76	10		2.0		12.0
Leather	60	9	12	10.0	0.4	8.6
Misc. organics	49	6	38	2.0	0.3	4.7
Dirt, ashes, etc.	25	3	1	0.5	0.2	70.3

Solution

Calculations are then done on the basis of say 100 kg of waste as shown in Table 3.3
Basis: 100 kg of waste
Empirical formula on dry basis:

$$C_{561}\,H_{906}\,O_{403}\,N_{6}\,S$$

$$\text{Moisture content} = \frac{100-82.6}{100} \times 100 = 17.4\%$$

TABLE 3.3
Sample Calculations

Component	Mass %	Moisture Content	Dry Mass (kg)	C	H	O	N	S	Ash
Newspaper	15	6	14.1	6.204	0.846	6.204	0.042	0.028	0.776
Other papers	24	5	22.8	10.032	1.368	10.032	0.068	0.046	1.254
Cardboard	33	5	31.35	13.794	1.881	13.794	0.094	0.063	1.724
Glass	4.2	0.5	4.18						
Plastic	0.48	2	0.48	0.288	0.034	0.11			0.048
Yard Wastes	17.97	60	7.19	3.45	0.431	2.731	0.216	0.022	0.338
Food Wastes	1.67	60	0.67	0.334	0.04	0.254	0.02	0.003	0.017
Dirt	2.01	8	1.85	0.462	0.055	0.018	0.009	0.004	1.3
Total			82.61	34.56	4.66	33.14	0.45	0.16	5.46
#moles				2.88	4.655	2.072	0.032	0.005	
Reduced Moles				561	906	403	6	1	

3.4 WASTE PROCESSING: PHYSICAL PROCESSES

3.4.1 Shredding

Shredding is usually the first step in solid waste handling. Shredding is used for particle size reduction. Shredding is achieved through the use of a flail mill, shaft hammer mill, and shear shredder, among others (Pfeffer, 1992). Shredding has been reported to be an energy-intensive step with a correlation between energy use and end-product particle size. The energy requirements for shredding can be estimated from equation 3.5.

$$\frac{dE}{dL} = -CL^{-n} \quad (3.5)$$

which states that the energy *dE* required to achieve a small size change *dL* in a unit mass of material is inversely proportional to the size of the article *L* (Worrell & Vesilind, 2010). If n=1,

$$E = C\log\frac{L_1}{L_2} \quad (3.6)$$

which is the Ficks law. E is the work done to bring the particle size from L_1 to L_2. If n=2, then the integrated form becomes:

$$E = C\left(\frac{1}{L_2} - \frac{1}{L_1}\right) \quad (3.7)$$

which is known as the Rittinger's law. Rittinger's law is used more for rough grinding operations, whereas Kick's law is better for fine grinding (Worrell & Vesilind, 2010).

3.4.2 Screening

Screening is a separation technique based on the particle size of the waste. Screens can be used after shredding. There are various screens that can be used for this purpose and they include rotary screens and vibratory screens. The rotary drum screen (trommel screen) consists of a large drum, which is usually elevated at an incline as shown in Figure 3.1.

The waste is usually added from the elevated size. The particles are carried to from the sides of the drum until they reach the top, where they then fall to the bottom of the drum. It is during the fall that the separation takes place. There are three solid waste behaviours inside a trommel screen, and these are cascading, cataracting, and centrifuging (Pichtel, 2005). Cascading occurs when the speed of the trommel screen is slow. If the speed is increased, then solid waste will rise to the top and slide backwards, leading to what is known as cataracting. At even higher speeds, the particles seem to be attached to the ides of the trommel leading to a phenomenon called centrifuging. The separation is improved with an increase in the number of rotations. The critical speed for a trommel screen is given by equation 3.8.

$$n_c = \left(\frac{g}{4\pi^2 r}\right)^{0.5} \tag{3.8}$$

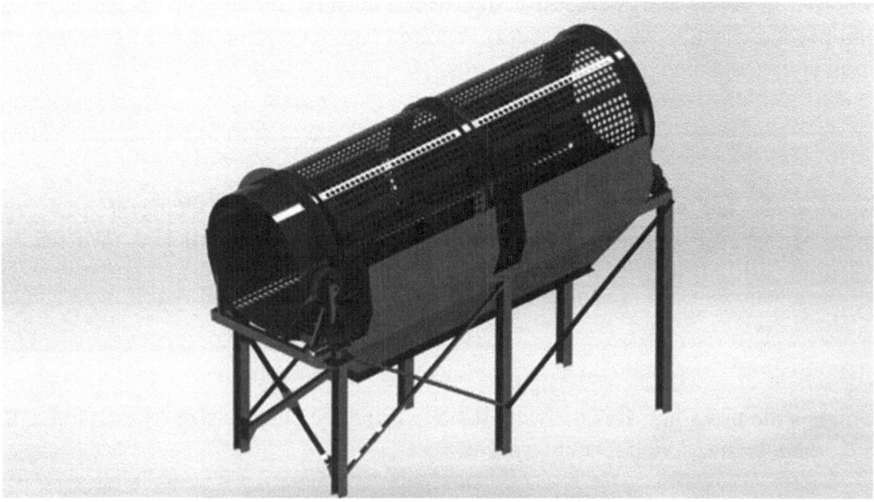

FIGURE 3.1 Trommel Screen *(https://grabcad.com/library/rotary-trommel-screen-1).*

where n_c is the critical speed (rotations/sec), g is the acceleration due to gravity (cm/sec^2), and r is the radius of trommel (cm) (Pichtel, 2005). Alternatively, equation 3.9 can be used to determine the size of the drum.

$$r\omega^2 = 32 \tag{3.9}$$

where r is the radius of drum and ω is the rotational speed in rad/sec. The throughput for a trommel can be calculated from equation 3.10.

$$D = \left(\frac{11.36Q_m}{d_b F k_v g^{0.5} \tan\alpha} \right)^{0.4} \tag{3.10}$$

where D is the trommel diameter (m), Qm is the trommel throughput, (kg/s), d_b is the bulk specific weight of solid waste (kg/m^3), α is the angle of trommel from base frame (deg), Kv is the velocity correction factor (Kv=1.35 when $\alpha = 3°$, and Kv=1.85 when $\alpha = 5°$), F is the fillage factor (a typical range is between 0.25 and 0.33), and g m/s^2 (Pichtel, 2005).

The calculation for screen recovery is given in equation 3.11 (Worrell & Vesilind, 2010) as follows:

$$R_{x1} = \left(\frac{x_1}{x_0} \right) \times 100 \tag{3.11}$$

Where R_{x1} is the screen recovery, %, x_1 is amount of material recovery as extract, that is, the material falling through the holes, mass/time, and x_0 is the amount of undersized material that *could have* fallen through the holes mass/time.

Example 3.2

Calculate the critical speed of a 2.9 m trommel screen.

Solution:

$$n_c = \left(\frac{g}{4\pi^2 r} \right)^{0.5}$$

$$n_c = \left(\frac{980}{4 \times 3.14 \times 3.14 \times 145} \right)^{0.5} = 0.414 \text{ rotations / s}$$

3.4.3 Magnetic Separation

This is a method to recover or separate ferrous components of the solid waste from non-ferrous components. Electromagnets or permanent magnets can be used. The efficiency of magnetic separation is affected by the magnetic force, speed of the conveyer belt, and the distance of the magnet from the conveyer belt. The closer the magnet is to the conveyor belt, the better the separation. The greater the magnetic force, the better the separation. High feed belt speed has been shown to be associated with lower recovery (Worrell & Vesilind, 2010).

3.5 THERMAL PROCESSES

3.5.1 Combustion

Combustion is a chemical reaction between oxygen and combustible material, in this case, the solid waste. Usually, air is supplied in excess to allow complete combustion. The theoretical amount of oxygen needed for complete combustion is called stoichiometric oxygen. Owing to suboptimal mixing of waste and gas within the combustion chamber, typically 1.8 to 2.5 times amount of stoichiometric oxygen is fed into the combustion chamber (Burnley, 2014). This is one of the cheapest methods of waste handling but suffers from uncontrolled environmental pollution. Grate firing is the most common technology for combustion (Leckner & Lind, 2020). Co-firing of solid waste with coal improves the combustion of solid waste (Zhu et al., 2022). There are several equations associated with incineration and combustion and some of them are given below:

$$C + O_2 \rightarrow CO_2$$

$$H_2 + 0.5O_2 \rightarrow H_2O$$

$$S + O_2 \rightarrow SO_2$$

$$2CO + O_2 \rightarrow 2CO_2$$

$$N_2 + O_2 \rightarrow 2NO$$

$$N_2 + 2O_2 \rightarrow 2NO_2$$

$$N_2 + 3O_2 \rightarrow 2NO_3$$

$$Cl + 2H_2O \rightarrow 4HCl + O_2$$

Example 3.3

Propylene (C_3H_6) is burned with 150% theoretical air. Given that the combustion is complete; air consists of 21% O_2 and 79% N_2 and R = 8314.34 J/kg-mole·K.

Calculate

(a) The air/fuel molar ratio.
(b) The molar analysis of the products of combustion.
(c) The volume of air required to burn gaseous propane to completion, if the ambient air temperature is 25°C.

Solution:

The combustion equation is given by:

$$2C_3H_6 + 9O_2 \rightarrow 6H_2O + 6CO_2$$

Basis of calculation: 1 mole of C_3H_6. From the equation, the amount of O_2 required for the complete combustion of 1 mole of C_3H_6 is 4.5 moles of O_2.

The amount of theoretical air required for 4.5 moles of O_2: 4.5/0.21 = 21.4 moles of air

The amount required for 150% theoretical air: (150/100) × 21.4 moles of air = 32.1 moles of air

The air/fuel molar ratio is given by 32.1/1 = 32.1 moles of air/mole of fuel.

(b) Amount N_2: 32.1×0.79 = 25.4 moles of N_2
A component table of the combustion process is shown below:

Component	Moles In	Moles Out	Mole Fraction in stack gas
C_3H_6	1	0	0
O_2	6.74	2.24	0.067
N_2	25.4	25.4	0.755
CO_2	0	3	0.089
H_2O	0	3	0.089
Total		33.64	1

(c) $$V = \frac{RT}{PM} = \frac{8314.34 \times 298.15}{101 \times 10^3 \times 29} = \frac{0.846 m^3}{kg} \text{ of air}$$

3.5.2 Incineration

Compared to other thermal processes, incineration can process large volumes of solid waste without the need for classification of the waste (Xing et al., 2021). The excess air used during incineration is dependent on the hearing value of the waste and the moisture content of the waste. The operating temperature of incinerators is between 900°C and 1100°C. Incineration suffers from the production of incineration ash, which becomes a source of heavy metal pollution (Chen et al., 2022). Incineration has a high volume reduction capability together with potential energy recovery (Liu et al., 2023). Incineration offers several advantages, including the non-sterile nature of ash residue and the potential reduction in the cost of operation through heat recovery techniques. However, the disadvantages include high initial capital costs, the need for skilled labour, and the fact that not all waste can be incinerated (Keith). There are various types of incinerators and these include open burning, single-chamber incinerators, tepee burners, open-pit incinerators, multiple-chamber incinerators, controlled air incinerators, central-station disposal, and rotary kiln incinerators. The heat from the burning of the waste is used to boil steam that drives turbines and in turn produces electricity.

3.5.3 Pyrolysis

Pyrolysis is a process where solid waste is exposed to elevated temperatures (300°C to 1300°C) in the absence of air (Koul et al., 2022). Pyrolysis processes can be classified as shown in Figure 3.2 based on operational parameters (Schmal, 2014; Kazawadi et al., 2021). The type of reactor used for each of these processes is also shown in Figure 3.2. Pyrolysis is useful for pathogen destruction and volume reduction. The liquid products derived from municipal solid waste typically contain 15–25% water, while the remaining organic fraction comprises acetic and other acids, sugars, and minor amounts of aldehydes, ketones, and alcohols. Pyrolysis takes place in sealed

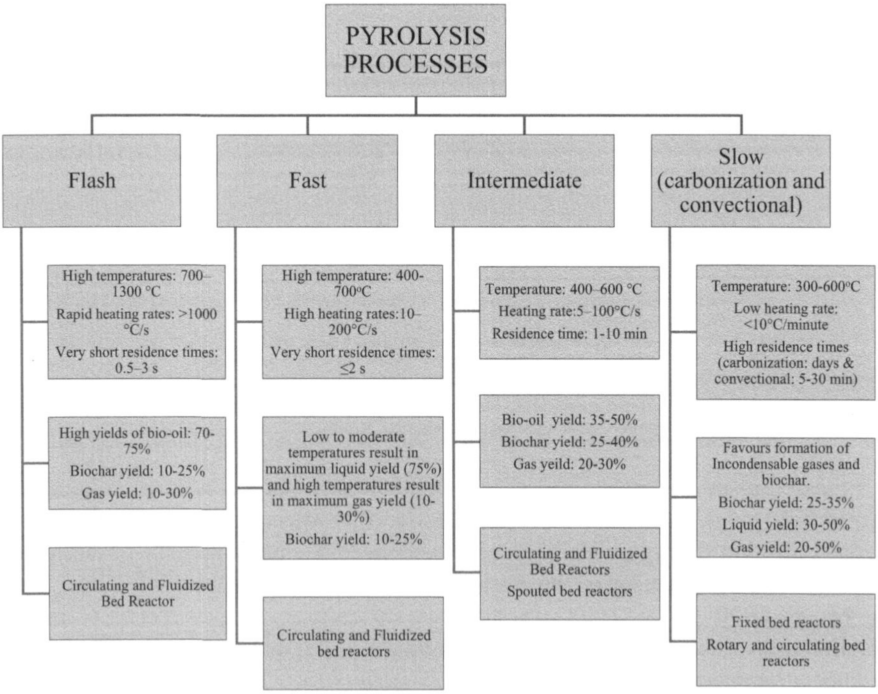

FIGURE 3.2 Classification of pyrolysis processes.

chambers although complete elimination of air is impossible. The products from pyrolysis are highly dependent on the operational conditions. Char is favoured by long exposure at temperatures in the range of 400 °C to 500°C. Short exposure at 500 °C to 1000 °C will result in more gas or liquid pyrolysis products. Rapid quenching of gas product is desired to increase the liquid product as compared to the gas product. Plasma pyrolysis vitrification is the preferred method for toxic waste. Pyrolysis is a highly endothermic process and can also be termed as destructive distillation. The products of pyrolysis include gases such as hydrogen, methane, carbon monoxide, and carbon dioxide; liquid products such as tar and oil; and char. Pyrolysis differs from gasification in that waste is heated in an externally enclosed unit. Due to the absence of air, the pyrolysis syngas is not high in nitrogen leading to a potentially higher calorific value as compared to the gasification syngas (Burnley, 2014).

3.5.4 Gasification

Gasification is a thermochemical process that transforms organic materials into fuel gas, thereby valorizing the solid waste (Robaina et al., 2021). Gasification is thought to happen in three stages and these are:

- Drying of solid feed to remove water
- Pyrolysis of waste to produce char, syngas, and hydrocarbons
- Gasification of solid char to produce mire syngas

Using a controlled temperature process gasification outputs include producer gas or syngas. Gasification process can be classified according to the heat source, where direct gasifiers use heat from partial combustion of feedstock, and a heat exchanger provides the heat for the thermal gasifier (Mishra et al., 2022). There are several parameters that affect gasification. Steam cracking is favoured by high temperatures while the heating value of the produced is favoured by low temperatures. High pressures improve efficiency but have a downside of being expensive to construct and operate. A smaller particle size of the feedlot favours an efficient gasification process. There are various types of reactors for gasification and these include downdraft, fixed bed, fluidized bed, drop tank, and fixed bed (Mishra et al., 2022). The syngas can then be used for chemical manufacture or energy generation affording the potential sustainability for the process. The so-called gasifying agents are used to partially oxidize the waste to produce a combustible gas and they include oxygen, air, and steam (Qin et al., 2022). Gasification using steam produces syngas that has low content of nitrogen, sulphur, and oxides (Mondal, 2022). Steam gasification produces high-calorie value syngas. The principal reactions in gasification are provided in the following equations (Sajid et al., 2022):

$$C + CO_2 \leftrightarrow 2CO$$

$$CH_4 + CO_2 \rightarrow 2CO + H_2$$

$$C + 2H_2 \rightarrow CH_4$$

$$H_2 + 0.5O_2 \rightarrow H_2O$$

$$C + 0.5O_2 \rightarrow CO$$

$$C + O_2 \rightarrow CO_2$$

$$C + H_2O \rightarrow CO + H_2$$

$$C + 2H_2O \rightarrow CO_2 + 2H_2$$

$$CH_4 + H_2O \rightarrow CO + 3H_2$$

$$CO_2 + H_2 \rightarrow CO + H_2O$$

Co-gasification is used to avoid the challenges of single-feed gasification, which include high-moisture content of municipal waste and seasonal availability of biomass (Mishra et al., 2022). The first step in gasification is pyrolysis where carbon-based materials are changed into char (Mishra et al., 2022). The produced carbon dioxide and water are reacted with char to form syngas. It has been shown that grapefruit gasification has the potential to produce 3.49 GJ of thermal energy per ton of GSW gasified using a fluidized bed gasifier (Robaina et al., 2021). The syngas so produced can be burnt without modification in a boiler to produce steam for electricity generation or it can be cleaned to remove impurities and moisture and then used in a turbine or gas engine.

3.6 BIOLOGICAL PROCESSES

3.6.1 Composting

Composting is an aerobic biochemical process that converts organic waste into manure. Bacteria, actinomycetes, fungi, worms, protozoa, and some larvae play an important role in composting. Composting results in the reduction of solid waste weight, volume, odour while increasing plant nutrient potential for the composted waste (Kumar, 2016). Composting can be categorized into a number of phases, including the lag phase, the exponential growth phase, and the maturation phase (Tchobanoglous & Kreith, 2002). The lag phase is when the composting conditions are established. The microbes are adjusting and getting used to the new environment. The length of the period is highly dependent on the type of substrate used in the solid waste. The lag period is very short when highly putrescible substrate is used, while it becomes longer with mixed solid waste and the period is very long when waste like dry waste is used (Tchobanoglous & Kreith, 2002). The active phase is associated with exponential growth in the number of microbes. It is associated with a temperature rise in the composting mass and the period is affected by obtaining conditions and type of substrate. The maturation phase is when the available substrate has been depleted and is associated with a temperature decline towards ambient conditions (Tchobanoglous & Kreith, 2002). There are various factors that affect composting. These include the carbon-to-nitrogen ratio (C:N), rate of aeration, moisture content, temperature, and pH. A C:N of 25:1 to 35:1 is preferred as a higher ratio results in the reduction of the decomposition rate while a lower ratio results in low energy supply, resulting in lower decomposition rates too (Koul et al., 2022). The usable oxygen should be 5% greater than the stoichiometric quantity, while pH has to be regulated to between 5.5 and 8.5. Single-component composting suffers from high-moisture content, low degradation rate, low ratio of carbon to nitrogen, small particle size, and high viscosity (Gao et al., 2021), which can be overcome by having multi-component substrate composting. The smaller the particle size, the faster the composting process although the 13 mm to 50 mm is termed the optimum range. Composting can also be an effective and cheap solid waste management system for remote rural communities (Gao et al., 2021). The benefits of composting include reduction in solid waste quantities and reduction in pathogenic organisms.

There are broadly three types of composting and these are windrows, in-vessel systems, and vermicomposting (Municipal solid waste in developing countries). Windrows is the least expensive system and makes use of windrows. It is suitable for use for garden waste. There are two types of windrows which are namely static and turned. Aeration for static composting is achieved without disturbing the windrow, while in turn, the aeration is achieved by demolishing the windrow. In-vessel composting is done in a container or building. This method is used for large volumes of waste and the container is designed in such a way to create adequate conditions for micro-organisms to thrive. The odours are controlled through the use of a proper ventilation system and a high C:N ratio (Kumar, 2016). Vermicomposting is the use of red earthworms to produce compost that can be used as mulch. It is particularly useful for food, paper, and yard waste. It is also important to note that vermicomposting does not generate high temperatures to kill pathogens (Kumar, 2016)

Composting calculations can be derived from mass balance as shown in Example 3.4.

Example 3.4

A poultry manure has a moisture content of 70% and 6.3% N (on dry basis). The manure is to be composted with oat straw with a moisture content of 20%. The desired C:N for the mixture is 25. The C:N ratio (dry mass basis) for poultry manure and oat straw is 15 and 48, respectively. The per cent nitrogen (dry basis) for oat straw is 1.1%. Determine the kg of oat straw required per kg of poultry manure to attain the desired C:N.

Solution:

Assumption

- 1 kg of moist poultry manure and X kg of moist oat straw are blended to achieve the desired C:N ratio

Calculations

- Dry mass of N_2 from poultry manure =1× (1–0.7) × 0.063 = 0.0189 kg
- Dry mass of C from poultry manure =1× (1–0.7)× 0.063×15 = 0.2835 kg
- Dry mass of N_2 from oat straw = X × (1–0.2)× 0.011 = 0.0088X kg
- Dry mass of C from oat straw = X × (1–0.2)× 0.011× 48 = 0.4224X kg

$$25 = \frac{\text{mass of C (poultry)} + \text{mass of C (manure)}}{\text{mass of N (poultry)} + \text{mass of N (manure)}}$$

$$25 = \frac{0.2385 + 0.4224X}{0.0189 + 0.0088X}$$

$$X = 1.16 \text{ kg}$$

The right amount of moisture for composting can be calculated using equation 3.12.

$$M_p = \frac{M_a X_a + 100 X_s}{X_s + X_a} \tag{3.12}$$

where
Mp is moisture in the mixed pile ready to begin composting, as percent moisture
Ma is moisture in the solids such as the shredded and screened refuse, as percent moisture
Xa is the mass of solids, wet tons
Xs is the mass of sludge or other sources of water, tons (This assumes that the solids content of the sludge is very low, a good assumption if waste activated sludge is used, which is commonly less than 1% solids.)

Example 3.5 (Adapted from Worrel & Vesilind, 2010)

Ten tons of a mixture of paper and other compostable materials has a moisture content of 7%. The intent is to make a mixture for composting of 50% moisture. How many tons of water or sludge must be added to the solids to achieve this moisture concentration in the compost pile?

Solution:

$$M_p = \frac{M_a X_a + 100 X_s}{X_s + X_a} \rightarrow 50 = \frac{(100 \times 7) + (100 X_s)}{X_s + 10} \rightarrow X_s = 50$$

The carbon and nitrogen levels can be calculated using equation 3.14.

$$C_p = \frac{C_r X_r + C_s X_s}{X_r + X_s} \tag{3.14}$$

where
Cp is the carbon concentration in the mixture prior to composting, as per cent of total wet mass of mixture
Cr is the carbon concentration in the refuse, as per cent of total wet refuse mass

Cs is carbon concentration in the sludge, as per cent of total wet sludge mass
Xs is the total mass of sludge, wet tons per day
Xr is the total mass of refuse, wet tons per day

3.6.2 Anaerobic Digestion

Anaerobic digestion (AD) is usually used for low solids substrate. It is a process where acetogens and methanogens are used to carry out hydrolysis, acidogenesis, acetogenesis, and methanogenesis (Jiang et al., 2022). Hydrolysis is the process where micro-organisms release hydrolytic enzymes, which break down complex biopolymers into simpler monomers or oligomers such as glucose, glycerol, and amino acids. Acidogenesis is continuation of hydrolysis where acids like acetic acid, butyric acid, and propionic acid use fermentative bacteria. Acetogenesis is the production of carbon dioxide, hydrogen, and acetate. Methanogenesis is the production of methane and is usually the slowest step, occurring simultaneously with acetogenesis. The process results in the production of methane and carbon dioxide. One of the advantages of anaerobic digestion is that it provides ways to utilize the nutrients for farming purposes (Pramanik, 2022). The process is affected by the type of substrate and the micro-organisms used. It is also worth noting that a substrate that is high in protein content can lead to production of toxic substances like ammonia (Pramanik, 2022). This can be averted by using a two-step AD process where hydrolysis-acidogenesis is separated from methanogenesis, leading to improved biogas production (Jiang et al., 2022). The development of technology has led to co-digestion in AD where two complementary substrates are used. A number of studies have used a combination of food waste with cattle manure, poultry manure, chicken manure, sewage sludge, and garden waste (Pramanik, 2022). The C:N ratio for some of these studies ranged from 13 to 41. Co-digestion has also allowed high solids substrates to be used for the production of methane using AD (Bao et al., 2023). In a study by Bao et al. (2023), a yield of 509.5 mLN/gVS was achieved with a 20% solids substrate of food waste:barley ratio of 5:1 (Bao et al., 2023). Methane production has also been shown to improve with the use of powdered activated carbon by reducing acidification under high organic loading rates (Li et al., 2022). Low-rate digesters take about 30–60 days to complete the digestion, while high-rate digesters take about 10–20 days. The process can occur at three temperature regimes and these are psychrophilic (<25°C), mesophilic (25°C–45°C) and thermophilic (45°C–70°C).

The reaction can be represented by equation 3.15.

$$C_aH_bO_cN_d + \left(4a - b - 2c + \frac{3d}{4}\right)H_2O \rightarrow \left(\frac{4a + b - 2c - bd}{8}\right)CH_4 + \left(\frac{4a - b + 2c + 3d}{8}\right)CO_2 + dNH_3 \quad (3.15)$$

The desired products are methane and sludge (Pichtel, 2005). The equation is useful in calculations of anaerobic digestion based on specific substrate as shown in example 3.6.

Example 3.6

Estimate the amount of CO_2 and CH_4 during anaerobic decomposition of glucose.

Solution:

Using equation 3.15

$$C_6H_{12}O_6 \rightarrow 3CH_4 + 3CO_2$$

Therefore, 180 kg of glucose will produce 48 kg of methane and 132 kg of carbon dioxide.

3.7 LANDFILLING

Landfilling is considered the most economical method for waste management (Saravanan et al., 2022). The most used landfilling methods are excavated cell/trench, area, and canyon (Keith). The excavated method is suitable for areas with adequate depth cover and where the water table is very deep. The excavated soil is used for covering the landfilled sites. This method allows for the progressive filling of the "cells" which also allows rehabilitation of the filled sites (Solid Waste Management). Lining must be used to prevent seepage and a variety of materials are useful for this purpose including geosynthetic products and low permeability clay (>1×10^{-9} ms^{-1}). Commercial bentonite can also be used as a lining material. Natural clay can be fractures and used as lining material because certain organic chemicals in waste can degrade clay integrity with time (municipal solid waste). High-density polyethylene (HDPE) can also be used liners. A combination of clay and HDPE can also be used for the construction of a liner. The area method is used in areas not suitable for excavation method. The solid waste is put into a mount on the soil surface and then covered by some earth. Lining of the surface is also critical to prevent seepage. The canyon methods make use of available borough pits and quarry pits as dump sites for solid waste. The downside of landfilling is the continued exposure by people who work at the landfill site. An improvement to this is the use of sanitary landfills which have engineering designs to collect the gases and the leachate emanating from the landfill. The base layer of a landfill must be designed with material which impermeable to protect groundwater (Parameswari et al., 2021). There are various factors that affect the location of a landfill and these include soil type, geology, and hydrogeology (Parameswari et al., 2021). There are numerous disadvantages of landfilling and these include uncontrolled release of gasses from the site, potentially uncontrolled seepage, and possible breeding of disease vectors due to poor control (Tchobanoglous & Kreith, 2002). Ground monitoring wells are a must to check if groundwater contamination is happening. Perforated pipelines located at the lowest points of the landfill are used to collect leachates. The leachates can be stored in containers at the landfill site where possible treatment can take place. The flow of the leachate through the liner is governed by Darcy's law as shown in equation 3.16.

$$Q = \frac{k \times A \times (h + d)}{d} \tag{3.16}$$

where Q is the leachate flow in m^3d^{-1}, k is the permeability of the liner in md^{-1}, A is the area of the site, h is the head of the leachate above the liner, and d is the thickness of the liner.

Example 3.7

Calculate the annual flow of leachate given the liner is 2 m and the permeability is 10^{-9} m d^{-1}, the leachate size is 3000 m^2, and the leachate head of leachate to 0.5 m.

Solution:

$$Q = \frac{k \times A \times (h+d)}{d} \times 365$$

$$Q = \frac{10^{-9} \times 3000 \times (2+0.5)}{2} \times 365 = 0.0014 m^3 / year$$

Estimation of methane formation:

The equation which can be used to estimate the methane generation of a landfill site is given as

$$Q_T = L_T \left(e^{kT}\left(1-e^{-kT}\right)\right) \tag{3.18}$$

QT is the amount of methane released in year T in tonnes, k is the rate constant in y^{-1},

$$L_T = DOC \times r_{mc} \times z \times \frac{16}{12} \tag{3.19}$$

DOC is the amount of degradable organic carbon in the original waste (in tonnes), *z* is the fraction of the *DOC* that degrades in the landfill, r_{mc} is the proportion of the degrading carbon that forms methane rather than carbon dioxide

3.8 ALTERNATIVE WASTE TREATMENT TECHNOLOGIES

The use of insects as an alternative waste management strategy for treating organic waste has gained traction in recent years due to the advantages it offers over convectional technologies. The most-studied insects are black soldier flies (*Hermetia illucens*), superworms (*Zophobas morio*), and mealworms (particularly the yellow *Tenebrio molitor*) (Kuan et al., 2022). The major advantages of insects are their ability to consume waste, transforming it into larval biomass and frass residue, which has found applications as animal protein in feedstocks, antimicrobial compounds, biofuels, and cosmetics (Cohn et al., 2024). The use of insects is one of the approaches to waste management that achieves a zero-waste circular economy by removing waste from the environment through insect consumption. This method keeps materials in the materials loop by transforming waste into valuable substrates. This is illustrated in Figure 3.2, which presents a circular economy view of organic waste management by insects. Studies on the use of insects have mainly focused on food waste and faecal sludge (secondarily sewage sludge), with only limited studies on other waste such as plastics. There is also a need to address other challenges with the use of insects, such as heavy metal accumulation, entomophobia, and economic viability concerns, since the technology requires higher investment and training, increased land use, availability of feedstocks, and proper blending ratios of waste to ensure consistent performance and product quality (Jones et al., 2024).

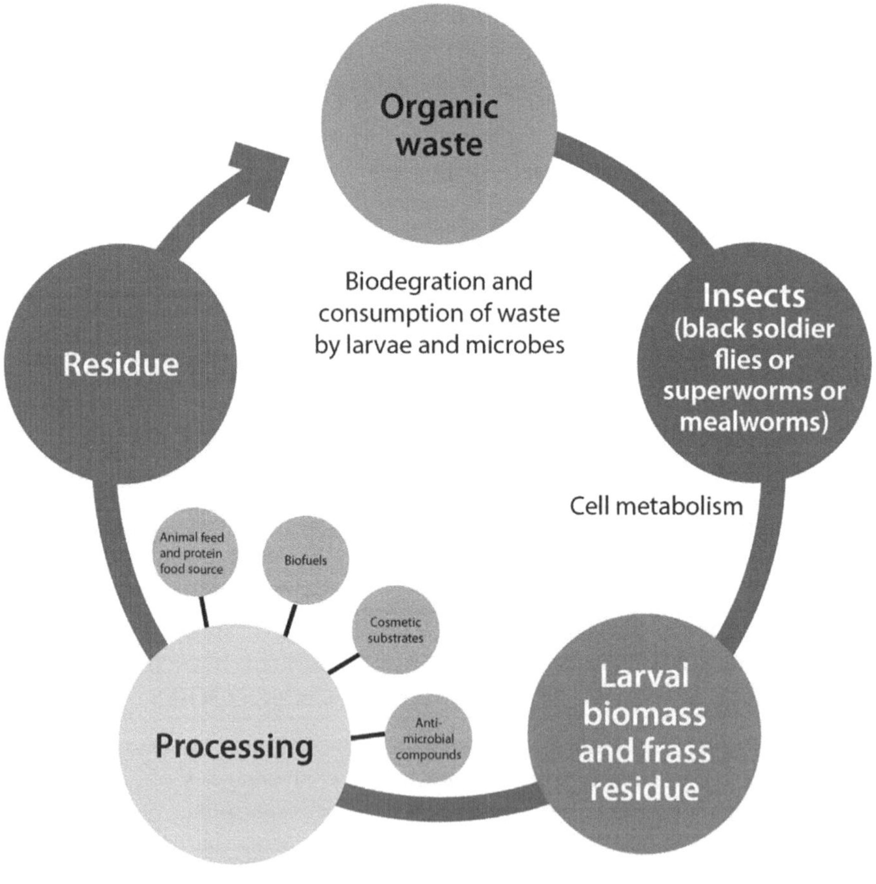

FIGURE 3.3 Use of insects to achieve a zero-waste circular economy.

CASE STUDY: CHICKEN LITTER PYROLYSIS IN BOTSWANA

The poultry industry in Botswana started in 1975 and has grown over the years making the country self-reliant in terms of poultry products – that is, meat and eggs. Poultry farmers in Botswana are characterized into two categories, namely traditional farming (family poultry) and commercial enterprise, which can be further categorized into large-scale (>50,000 birds), medium scale (20,001 to 50,000 birds) and small scale (≤20,000 birds) (Moreki, 2010). The rearing of chickens is an energy-intensive process requiring energy for feeding, lighting, heating, ventilation, and cooling with feeding accounting for about 55% of electricity consumption in automated birdhouses. In addition to this, poultry litter (PL) is produced as waste and comprises a mixture of chicken faeces, feathers, and bedding materials (wood shavings and sawdust). In Botswana, on average, 20 kg of PL is generated per broiler bird reared. Given this scenario, there is a need to develop a sustainable circular process to minimize energy usage and also manage the waste produced.

Among the technologies available for managing PL, pyrolysis was investigated as a possible waste management option in Botswana by Pyro Carbon Energy (PCE) using an induction heated auger-type reactor. The project was funded by the Botswana Innovation Fund (BIF) and resulted in a one-ton per day (1 TPD) commercial-scale proof-of-concept plant chicken litter pyrolysis plant (Figure 3.4). The main objective of the work was the thermochemical

FIGURE 3.4 3D model of the Containerized Pyrolysis Unit (CPU).

Source: Copyright PCE, pyrocarbonenergy.com.

treatment of PL using pyrolysis to generate valuable products such as biofuel and biochar coupled with cogeneration of electricity.

The main products obtained from the pyrolysis of chicken litter were biochar, bio-oil, and pyrolytic gas. Biochar can be used as a composting and soil enhancer, and also as an adsorbent for water treatment. Bio-oil produced has high caloric value (>40 MJ/kg) and has properties that make it suitable for blending with diesel or use as heavy fuel oil (HFO). Pyrolytic gas can be readily used in gas engines to produce electricity, or it can also be used as a heating fuel.

Based on this article and the background information given above, together with other relevant resources, discuss the following questions:

1. In what ways does the proposed approach (pyrolysis of PL) contribute to a sustainable circular economy?
2. What other options can you propose as alternatives to pyrolysis as a method of managing PL?
3. What are the drawbacks and advantages of the proposed options (part 2) when compared to pyrolysis of PL?

3.9 PROBLEMS

1. A poultry manure has a moisture content of 70% and is 6.3% N (on dry basis). The manure is to be composted with oat straw with a moisture content of 20%. The desired C:N for the mixture is 30. The C:N ratio (dry mass basis) for poultry manure and oat straw is 15 and 48, respectively. The % nitrogen (dry basis) for oat straw is 1.1%. Determine the quantity (kg) of oat straw required per kilogramme of poultry manure to attain the desired C:N.
2. Ideally, the production of methane and carbon dioxide under anaerobic conditions can be represented by the following equation:

$$C_aH_bO_cN_d + \left(\frac{4a - b - 2c + 3d}{4}\right)H_2O \rightarrow \left(\frac{4a + b - 2c - 3d}{8}\right)CH_4 + \left(\frac{4a - b + 2c + 3d}{8}\right)CO_2 + dNH_3$$

A rough estimate of MSW composition is as follows:

Component	Weight %
Paper (assume all cellulose, $C_6H_{10}O_5$)	50
Glass	10
Steel	10
Aluminium	10
Food waste (assume it is sugar, $C_6H_{12}O_6$)	20

2.1 If 10,000 tonnes of this material is placed in a landfill, how much methane gas would theoretically be produced by anaerobic decomposition?
2.2 What is the methane production per kilogram of waste? Comment on the answer.
2.3 What are two good uses for methane collected from a landfill?

3. Determine the power required to shred 100 ton/day of solid waste having an 80th percentile size of 30 cm to a product having an 80th percentile size of 1.7 cm. Assume that the typical bond work index Wi = 400 kWh/ton.
4. The light fraction components of a solid waste after shredding and air classification are given below:

Component	Mass %	% Moisture Content
Food waste	16.00	60
Paper	40.00	6
Cardboard	8.00	5
Plastics	12.00	2
Textiles	7.00	10
Rubber	4.50	2
Leather	4.50	9
Glass	8.00	0.5

Determine the empirical formula of the waste on a dry basis using the information given below:

Molar masses: C = 12; H=1; O = 16; N =14; S = 32

	% by mass (dry basis)					
Ultimate analysis	**C**	**H**	**O**	**N**	**S**	**Ash**
Yard wastes	48	6	38	3.0	0.3	4.7
Wood	50	6	43	0.2	0.1	0.7
Food wastes	50	6	38	3.0	0.4	2.6
Paper	44	6	44	0.3	0.2	5.5
Cardboard	44	6	44	0.3	0.2	5.5
Plastic	60	7	23			10.0
Textiles	56	7	30	5.0	0.2	1.8
Rubber	76	10		2.0		12.0
Leather	60	9	12	10.0	0.4	8.6
Misc. organics	49	6	38	2.0	0.3	4.7
Dirt, ashes, etc.	25	3	1	0.5	0.2	70.3

REFERENCES

ASTM D5231-92. (2003). *Standard test method for determination of the composition of unprocessed municipal solid waste* (p. 6). ASTM International. https://store.astm.org/d5231-92r16.html

Bao, R., Wei, Y., Guan, R., Li, X., Lu, X., Rong, S., Zuo, X., & Yuan, H. (2023). High-solids anaerobic co-digestion performances and microbial community dynamics in co-digestion of different mixing ratios with food waste and highland barley straw. *Energy*, *262*(PB), 125529. https://doi.org/10.1016/j.energy.2022.125529

Berge, S., & von Blottnitz, H. (2022). An estimate of construction and demolition waste quantities and composition expected in South Africa. *South African Journal of Science*, *118*, 1–5. https://doi.org/10.17159/sajs.2022/12485

Burnley, S. (2014). Solid management. *Book*, 1–336.

Chen, M., Oshita, K., Takaoka, M., & Shiota, K. (2022, April). Co-incineration effect of sewage sludge and municipal solid waste on the behavior of heavy metals by phosphorus. *Waste Management*, *152*, 112–117. https://doi.org/10.1016/j.wasman.2022.08.010

Cohn, Z., Latty, T., Abbas, A., Jones, A. K., Nur-Aliah, N. A., Ivorra, T., & Heo, C. C. (2024). Black soldier fly (Diptera: Stratiomyidae) reduction of different sludges, subsequent safety, and research gaps. *Journal of Environmental Management*, *137*(September 2021), 119394. https://doi.org/10.1016/j.wasman.2021.10.013

Crosset, E. (2014). *Opportunities and challenges of a sustainable solid waste management in Tsumeb*. Royal Institute of Technology.

Gao, Y., Tan, L., Liu, F., Li, Q., Wei, X., Liu, L., Li, H., Zheng, X., & Xu, Y. (2021, July). Optimization of the proportion of multi-component rural solid wastes in mixed composting using a simplex centroid design. *Bioresource Technology*, *341*, 125746. https://doi.org/10.1016/j.biortech.2021.125746

Hassan, S. H., Aziz, H. A., Daud, N. M., Keria, R., Noor, S. M., Johari, I., & Shah, S. M. R. (2018). The methods of waste quantification in the construction sites (A review). *AIP Conference Proceedings*, *2020*(October 2018). https://doi.org/10.1063/1.5062682

Jiang, H., Wang, S., Li, B., Feng, L., Zhai, L., Zhou, H., Li, Y., & Pan, J. (2022, June). Anaerobic digestion of organic fraction of municipal solid waste using a novel two-stage solid-liquid system. *Journal of Cleaner Production*, *370*, 133521. https://doi.org/10.1016/j.jclepro.2022.133521

Jones, A. K., Nur-Aliah, N. A., Ivorra, T., & Heo, C. C. (2024). Black soldier fly (Diptera: Stratiomyidae) reduction of different sludges, subsequent safety, and research gaps. *Journal of Environmental Management*, *349*(October 2023), 119394. https://doi.org/10.1016/j.jenvman.2023.119394

Kazawadi, D., Ntalikwa, J., & Kombe, G. (2021). A review of intermediate pyrolysis as a technology of biomass conversion for coproduction of biooil and adsorption biochar. *Journal of Renewable Energy*, *2021*, 1–10. https://doi.org/10.1155/2021/5533780

Koul, B., Yakoob, M., & Shah, M. P. (2022). Agricultural waste management strategies for environmental sustainability. *Environmental Research*, *206*, 112285.

Kuan, Z. J., Chan, B. K. N., & Gan, S. K. E. (2022). Worming the circular economy for biowaste and plastics: Hermetia illucens, Tenebrio molitor, and Zophobas morio. *Sustainability (Switzerland)*, *14*(3), 1–13. https://doi.org/10.3390/su14031594

Kumar, S. (2016). Municipal solid waste management in developing countries. *Municipal Solid Waste Management in Developing Countries*. https://doi.org/10.1201/9781315369457

Leckner, B., & Lind, F. (2020). Combustion of municipal solid waste in fluidized bed or on grate—a comparison. *Waste Management*, *109*, 94–108. https://doi.org/10.1016/j.wasman.2020.04.050

Li, L., Gao, Q., Liu, X., Zhao, Q., Wang, W., Wang, K., Zhou, H., & Jiang, J. (2022, March). Insights into high-solids anaerobic digestion of food waste enhanced by activated carbon via promoting direct interspecies electron transfer. *Bioresource Technology*, *351*, 127008. https://doi.org/10.1016/j.biortech.2022.127008

Liu, X., Duan, L., Duan, Y., Li, L., Sun, Z., & Sun, G. (2023). Improved fuel conversion through oxygen carrier aided combustion during incineration of biomass-based solid waste in a rotary kiln. *Fuel*, *331*(P1), 125714. https://doi.org/10.1016/j.fuel.2022.125714

Mishra, R., Singh, E., Kumar, A., Ghosh, A., Lo, S. L., & Kumar, S. (2022, September). Co-gasification of solid waste and its impact on final product yields. *Journal of Cleaner Production*, *374*, 133989. https://doi.org/10.1016/j.jclepro.2022.133989

Mondal, P. (2022). From municipal solid waste (MSW) to hydrogen: Performance optimization of a fixed bed gasifier using Box-Benkhen method. *International Journal of Hydrogen Energy*, *47*(46), 20064–20075. https://doi.org/10.1016/j.ijhydene.2022.04.150

Moreki, J. C. (2010). Opportunities and challenges for the Botswana poultry industry in the 21st century: A review. *Livestock Research for Rural Development*, *22*. http://www.lrrd.org/lrrd22/5/moreb22089.htm

Nakin, M., Kabiti, H., Ndhleve, S., & Tsheleza, V. (2022). Household solid waste quantification, characterization and management practices in Mthatha city, South Africa. *International Journal of Environment and Waste Management*, *1*(1), 1. https://doi.org/10.1504/ijewm.2022.10035404

Nell, C. M., Schenck, C., & Waal, J. de. (2022). Waste characterisation in Stellenbosch Local Municipality, South Africa. *South African Journal of Science*, *118*, 1–12.

Parameswari, K., Majid Salim Al Aamri, A., Gopalakrishnan, K., Arunachalam, S., Ali Said Al Alawi, A., & Sivasakthivel, T. (2021). Sustainable landfill design for effective municipal solid waste management for resource and energy recovery. *Materials Today: Proceedings*, *47*, 2441–2449. https://doi.org/10.1016/j.matpr.2021.04.528

Pfeffer, J. T. (1992). *Solid waste management engineering* (W. J. Hall, Ed., 1st ed.). Prentice Hall.

Pichtel, J. (2005). *Waste management practices municipal, hazardous, and industrial* (1st ed.). Taylor & Francis Group.

Pramanik, S. K. (2022, September). Anaerobic co-digestion of municipal organic solid waste: Achievements and perspective. *Bioresource Technology Reports*, *20*, 101284. https://doi.org/10.1016/j.biteb.2022.101284

Qin, J., Zhang, Y., Heberlein, S., Lisak, G., & Yi, Y. (2022, April). Characterization and comparison of gasification and incineration fly ashes generated from municipal solid waste in Singapore. *Waste Management*, *146*, 44–52. https://doi.org/10.1016/j.wasman.2022.04.041

Ramakrishna, V., Subrahmanyam, C. N., Bhanuchand, S., Rao, B. S., & Kumar, A. N. V. A. (2016). Municipal solid waste quantification, characterization and management in Mylavaram. *IOSR Journal of Mechanical and Civil Engineering*, *13*(5), 77–87. https://doi.org/10.9790/1684-1305047787

Robaina, B. A. R., Reyes, Y. G., Trujillo, L. A., Montesino, F. M., Pedroso, D. T., Machin, E. B., Machín, A. B., Pascual, R., Pérez, J. A., Callejo, A. G., & Cebrian, J. L. S. (2021, July). Assessment of fluidized bed gasification of grapefruit solid waste. *Bioresource Technology Reports*, *15*. https://doi.org/10.1016/j.biteb.2021.100782

Sajid, M., Raheem, A., Ullah, N., Asim, M., Ur Rehman, M. S., & Ali, N. (2022, April). Gasification of municipal solid waste: Progress, challenges, and prospects. *Renewable and Sustainable Energy Reviews*, *168*, 112815. https://doi.org/10.1016/j.rser.2022.112815

Saravanan, A., Kumar, P. S., Nhung, T. C., Ramesh, B., Srinivasan, S., & Rangasamy, G. (2022). A review on biological methodologies in municipal solid waste management and landfilling: Resource and energy recovery. *Chemosphere*, *309*(P1), 136630. https://doi.org/10.1016/j.chemosphere.2022.136630

Schmal, M. (2014). *Chemical reaction engineering: essentials, exercises and examples*. CRC Press, Taylor & Francis Group.

Tchobanoglous, G., & Kreith, F. (2002). *Handbook of solid waste management* (2nd ed.). The McGraw-Hill Companies.

Villoria-Sáez, P., Porras-Amores, C., & del Río Merino, M. (2020). Estimation of construction and demolition waste. In *Advances in construction and demolition waste recycling*. Elsevier Ltd. https://doi.org/10.1016/b978-0-12-819055-5.00002-4

Weichgrebe, D., Speier, C., & Mondal, M. M. (2017). Scientific approach for municipal solid waste characterization. In S. Goel (Ed.), *Advances in solid and hazardous waste management* (pp. 65–100). Springer International Publishing.

Worrell, W. A., & Vesilind, P. A. (2010). *Solid waste engineering* (2nd ed.). Global Engineering.

Xing, Z., Ping, Z., Xiqiang, Z., Zhanlong, S., Wenlong, W., Jing, S., & Yanpeng, M. (2021). Applicability of municipal solid waste incineration (MSWI) system integrated with pre-drying or torrefaction for flue gas waste heat recovery. *Energy*, *224*, 120157. https://doi.org/10.1016/j.energy.2021.120157

Zhu, S., Yang, Y., Yan, Y., Causone, F., Jin, X., Zhou, X., & Shi, X. (2022). Experimental study on combustion, ash fusibility and slagging propensity during co-combustion of organic solid waste and lignite. *Journal of the Energy Institute*, 101145. https://doi.org/10.1016/j.joei.2022.11.005

4 Environmental Management Legislation in Sub-Saharan Africa

Freeman Ntuli

4.1 HISTORICAL BACKGROUND

Pollution of the environment has been around for a long time, although one might not have been aware of it. Developments in human civilization, coupled with rapid population growth and urbanization, have greatly increased the demand for goods and services. This has put immense pressure on the industrial sector for greater productivity and continuous development of better and more efficient products. Since greater emphasis was put on productivity, little was done to assess the environmental impacts of many processes and products. The resultant transformation in the technology of production has been held responsible for nearly all the environmental degradation that has occurred in the twentieth century (Commoner, 1988). The transition from reusable packaging for most commonly used household products such as milk, soft drinks, and beer, which were traditionally packaged in reusable glass containers, to single-use plastic and can packaging has also resulted in a significant increase in the amount of waste disposed from households. This transition, coupled with rapid urbanization and a growing culture of consumerism in most developing countries, has resulted in a rapid increase in household waste which has outpaced the development of waste management infrastructure, leading to uncontrolled disposal of waste. Packaging is now considered one of the major consumers of raw materials with an estimated 40% usage in plastics and 50% usage in paper in Europe (Coelho et al., 2020). The estimated amount of waste generated from this packaging in Europe is about 36% (Eurostat, 2019) of the municipal solid waste and around 22% in sub-Saharan Africa (United Nations Environment Programme, 2018).

It is only recently that man has taken note of the negative impacts on the environment that have come about as a result of rapid industrialization. In 1983, the United Nations General Assembly established the World Commission on Environment and Development, intending to consider the linked issues of environment and development. It set out the conditions by which humanity could continue to enjoy the benefits of industrial development without creating further deterioration and damage to the environment. It was this commission that introduced the concept of sustainable development, which was later incorporated into Agenda 21 of the Rio Earth Summit in 2012 (UNCED, 1992). This has led to the introduction of environmental legislation by the local, national, and international communities to try and curb the

DOI: 10.1201/9781003377856-4

effects of environmental degradation. Key to this development is the drive towards the incorporation of the principles of the circular economy in the legislative policies in order to close the materials flow loop and meet the SDGs. Progress has been made in implementing the principles of the circular economy in a number of European Union (EU) regulations and policies and also in China, which was the first to adopt a law on the circular economy in 2008 (Korhonen et al., 2018). Other countries that have made significant progress in this direction include Japan, Sweden, the United Kingdom, Finland, the Netherlands, and Canada (Korhonen et al., 2018). Sub-Saharan African countries currently do not have adequate national policies that address the principles of the circular economy in their legislation (Debrah et al., 2022). However, a few countries that are members of the African Circular Economy Alliance (ACEA), which was formed in 2016, namely Rwanda, South Africa, Nigeria, Ghana, Côte d'Ivoire, Benin, Burkina Faso, and Sudan, are making progress in this direction.[1]

4.2 ENVIRONMENTAL LEGISLATION APPROACHES

Environmental legislation is basically composed of laws regulating human activities which are damaging or threatening to the environment. The objective of environmental laws must be remediation, not punishment. As a result, the government, industry, and society must be involved in pollution prevention. Environmental laws are implemented through regulations or standards that define specific pollutant limits for effluents discharged into the environment.

Environmental legislation is usually based on two approaches:

- BAT – best available technology
- BPT – best practicable (reasonable) technology

The BAT approach uses the best technology available on the world market for the development of processes and waste management practices.

The BAT approach implies uniform standards and does not take into account the assimilative properties of the receiving media. It is most suitable for well-industrialized and advanced countries where finance is not a limiting factor and where rivers and the environment are grossly polluted (Akuffo, 1998).

However, it is not recommended for developing countries because they cannot afford the cost of such technology and are usually unduly restrictive (Akuffo, 1998).

The BPT approach is based on the maintenance of a delicate balance between the following:

- The ability of industries, nation, and individuals to pay for pollution control facilities
- The maintenance of minimum standards of environmental quality in relation to the requirements of public health and the health of the environment.

It is the best approach for developing countries, as it avoids the pursuit of idealistic maximum standards inherent in BAT, which may cause economic dislocation in some sectors of the economy.

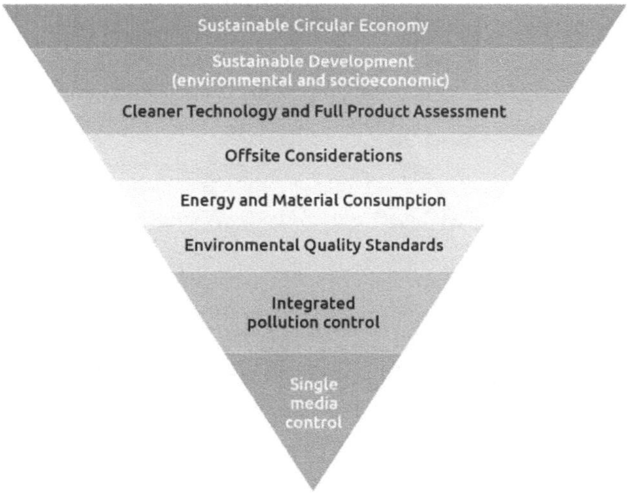

FIGURE 4.1 The widening scope of the pollution control law.

Currently, more emphasis is being put on pollution prevention rather than pollution abatement. The scope of industrial pollution legislation has been widening from a single media approach, through integrated pollution control (IPC), to integrated pollution prevention control (IPPC) and further towards sustainable development (Nicholas et al., 2000), as shown in Figure 4.1. This is reflected in the development of the environmental legislation in Europe and the United States. This has led to the introduction of the EC directive on IPPC (EC, 1996) and the Pollution Prevention Act (PPA, 1990) implemented by the United States Environmental Agency (USEPA). These regulations are consistent with international emphasis on sustainable development, which takes into account environmental and socioeconomic factors such as resource depletion.

4.3 REGULATORY FRAMEWORK STRUCTURE

Environmental laws and regulations are normally tailored to regulate activities that may impact the air, water, and environment around us. These laws are normally informed by a set of principles or guidelines, which are outlined below:

1. The precautionary principle: based on controlling activities suspected to cause environmental degradation in the absence of concrete scientific data
2. The prevention principle: the fundamental notion on which most environmental laws are based and is focused on preventing the discharge of pollutants into the environment.
3. The "polluter pays" principle: an economic model based on transferring the cost of pollution remediation to the generator of pollutants.
4. The integration principle: a holistic approach to environmental protection through the integration of sustainable development objectives in all developmental activities, laws, and regulations.

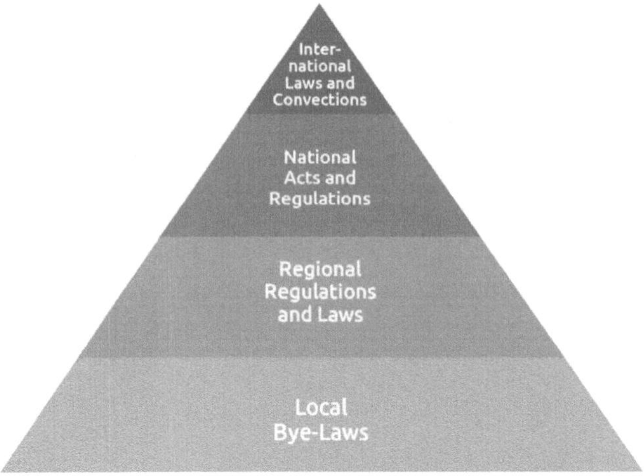

FIGURE 4.2 Legal framework for environmental laws.

5. The public participation principle: driven by citizen participation in public forums and legal lawsuits.
6. Sustainable development principles: based on regulating economic activity that preserves the quality of the environment for future generations.
7. Sustainable circular economy: based on minimizing the loss of materials and energy from the material and energy loop.

The legal framework for environmental laws is normally structured as shown in Figure 4.2.

Most of the regulatory control of environmental laws is carried out at the local level (district, municipal, council) through local bye-laws that are tailored to address national and regional regulations. The implementation and management of these laws is a government responsibility through local and regional government structures such as city and town councils.

4.4 ENVIRONMENTAL MANAGEMENT REGULATIONS AND LAWS

Management of emissions to the environment (water, soil, and air) is regulated by law in order to prevent environmental pollution and protect human and animal life. Even though this chapter will focus on the management of solid waste, it is critical to note that solid waste can generate secondary emissions into the air and water. These secondary emissions are governed by specific acts and regulations for air or water emissions. The section below will review acts, regulations, and bye-laws in a number of countries in sub-Saharan Africa in order to highlight similarities and differences. Representative countries have been chosen from the regional blocks within the SSA countries, namely Central Africa, East Africa, Southern Africa, and West Africa.

4.5 WEST AFRICA

4.5.1 Nigeria

Nigeria is a West African state and is the most populous country in Africa, with an estimated population of 225 million as of 2023. It occupies a land area of about 923,769 square kilometres, translating to a population density of about 243 people per square kilometre. Its capital city is Abuja. Lagos state is the smallest state in Nigeria; however, it has the highest urban population, accounting for about 27% of the national population (Lagos State Government official website).[2] It has a territorial land area of 3518.61 square kilometres and served as the state capital until 1991 and federal capital until 1976. Lagos is Nigeria's financial hub, and its gross domestic product (GDP) accounts for 26.7% of Nigeria's total GDP and more than 50% of non-oil GDP (Lagos State Government official website, https://lagosstate.gov.ng/).

The need to protect the environment is enshrined in the Constitution of the Federal Republic of Nigeria, 1999 (as amended),[3] in Section 20. It states that "The State shall protect and improve the environment and safeguard the water, air and land, forest and wildlife of Nigeria". The main national environmental laws are the Environmental Impact Assessment Act 1992,[4] Federal Environmental Protection Agency Act 1988,[5] and the National Environmental Standards and Regulations Enforcement Agency (Establishment) Act 2007.[6] The Environmental Impact Assessment Act requires that significant environmental issues be identified and studied before commencing or embarking on any project or activity that is likely to have adverse effects on the environment. The Federal Environmental Protection Agency has the responsibility of protecting the environment and promoting the sustainable use of Nigeria's natural resources by initiating policies related to environmental research and technology. The National Environmental Standards and Regulations Enforcement Agency (NESREA) is the enforcement agency for environmental standards, regulations, rules, laws, policies, and guidelines and provides oversight over federal agencies in the country.

The overarching law that operates in Lagos State is the Lagos State Environmental Protection Law[7] enforceable by Lagos State Environmental Protection Agency (LASEPA).[8] This law covers all aspects of environmental effluent discharge to the environment. The term effluent is defined as "waste generated from manufacturing service operations business", and the term environment is defined as "water, air, land and the interrelationship which exist among water, air, land and all living things and the medium of air includes the air within buildings and air without other natural or man-made structures above or below the ground". The law also deals with the licensing requirements, handling of effluent discharges, monitoring of environmental discharges, and penalties for non-compliance for both commercial and household sectors. The Law also empowers LASEPA to set emissions standards for effluent discharges to the environment and the criteria to control the pollution of the environment based on the BPT approach.

It should be noted that apart from this agency, there is Lagos State Waste Management Authority (LAWMA). LAWMA is the first waste management outfit in West Africa and is responsible for the collection and disposal of municipal and industrial waste in Lagos State. LAWMA has its own laws and regulations, however, LAWMA law is subject to federal and state law.

The national agency enforces waste management laws is NESREA, and this institution has its laws and regulations (National Environmental Standards and Regulations Enforcement Agency (Establishment) Act, 2007 2007 Act no. 25).[9]

The current laws are mainly based on integrated pollution control and environmental quality standards, with not much emphasis on higher considerations such as sustainable development and sustainable circular economy (Figure 4.1). The laws are mainly enforced and guided through the "polluter pays principle" and the "precautionary principle". There is currently no single, all-encompassing national law or regulation that addresses environmental protection and sustainability.

4.5.2 Gambia

Gambia is a West African state and is one of the smallest countries in Africa, with an estimated population of about 2.4 million as of 2023. Its capital is Banjul, and the size of the country is about 11,300 square kilometres. Its population density is about 212 people per square kilometre.

In the Gambia, as opposed to Nigeria, there are no district or federal laws but a national law that is enforceable across all the district/council areas. The overarching act for the control and management of the environment in Gambia is the National Environment Management Act, 1994.[10] The Act is administered by the National Environment Management Council, which is a ministerial body tasked with policy-making, administration, and coordination of environmental plans and policies. The management function of the Act is delegated to the National Environment Agency (NEA),[11] which serves as the Secretariat of the National Environment Management Council. The Agency is the principal body responsible for the management of the environment and coordination of all government efforts in the area of environmental management. The Agency's key role is to implement policies of the Council pertaining to the environment, create awareness of environmental issues, develop environmental guidelines, regulations, and standards, and carry out environmental audits and monitoring. The Act also makes provision for the establishment of local environmental committees in municipal areas, namely Banjul and Kanifing municipal areas, which act as organs of the Agency in implementing the Act in their respective territories. The Act also empowers the Agency to carry out environmental impact assessments for projects listed in Schedule A of the Act and approval of proposed projects before issuing a license or permit to operate and carry out periodic environmental audits on existing and new projects.

Management of emissions in the Act is through the use of environmental quality standards that specify the minimum levels allowable for emissions or discharge to the air, water, land, and the environment in general – that is, solid waste, noise, vibrations, and radiation. Some of the important regulations and standards implemented and by NEA[12] are:

- Environmental Protection (Prevention of Dumping) Act, 1988
- Hazardous Chemicals Regulations, 1999

- Environmental Discharge (permitting) Regulations 1999
- Environmental Quality Standards, 1999

The Act focuses more on single media control and does not adequately address issues of integrated pollution control and sustainability. The Act uses the "polluter pays principle" as a means of managing and controlling environmental pollution.

Additional legislation dealing with solid waste management includes the Waste Management Bill, 2003 (Waste Management Act, 2003),[13] which is regulated by the Waste Regulation Authority, designated as the "Agency". The Waste Management Board is responsible for the policy direction of the "Agency" as far as its role as a Waste Regulation Authority is concerned. The Act delegates the Municipal Council with the responsibility for waste collection and waste disposal authority within its locality and enforcement of the Anti-Littering Regulations, 2007.[14]

4.5.3 Ghana

Ghana is a West African state with its capital in Accra. It is the second-largest producer of gold and cocoa on the African continent. It also has reserves of oil and diamond. It has an estimated population of 32.1 million and occupies a land area of 238,535 square kilometres, translating to a population density of 135 people per square kilometre, which is far lower than Nigeria and Gambia.

The right to protect and safeguard the environment is enshrined in the Constitution of the Republic of Ghana (the Constitution of the Republic of Ghana, 1992 (as amended in 1996)).[15] In Ghana, the Environmental Protection (Act 490 of 1994) 106[16] established the authority, responsibility, structure, and financing of the Environmental Protection Agency (EPA). The EPA is the regulatory agency established to regulate and implement environmental laws and policies in Ghana. Act 490 amends and consolidates the laws relating to environmental protection, pesticide control, and other environmental issues. It also empowers the EPA through a legislative instrument (LI) in consultation with the relevant minister to make regulations for environmental protection, licence and issue environmental permits, and monitor environmental compliance.

The main environmental laws and policies in relation to waste and effluent management in Ghana are as follows:

- Environmental Protection Agency Act, 1994 (Act 490) Part I & II
- Ghana Water and Sewerage Corporation Act 1965 (Act 310)
- Water Resources Commission Act, 1996 (Act 522)
- Abandoned Property (Disposal) Act, 1974
- Environmental Assessment Regulations 1999, (LI 1652)
- Layout Designs, Local Government Act (1994), Act 462.

A review of Ghana's environmental laws and policies and their effectiveness has been presented by Mensah et al. (2020). Their findings indicate that the current laws and policies on environment are fragmented; there is need for capacity building in terms of personnel and data leveraging on technological advances; they are ineffective in addressing sustainable environmental development, and there is a lack of community awareness on the relation between human activity and the environment (Mensah et al., 2020).

The current laws and policies do not reflect sustainable circular economy principles and are more focused on single media control. The current legislation would need to be reviewed to address sustainable circular economy principles, as it is currently based more on the prevention and polluter pays principles.

4.6 SOUTHERN AFRICA

4.6.1 South Africa

South Africa is a Southern African state which has one of the biggest and most developed economies in Africa, with its executive, legislative, and judicial capitals in Pretoria, Cape Town, and Bloemfontein, respectively. It occupies an area of 1,221,037 square kilometres and has an estimated population of 60.6 million as of 2023. This translates to an estimated population density of 50 people per square kilometre. The population density is about four times lower than that of Nigeria and Gambia. Gauteng province is the smallest province by land area in South Africa, occupying approximately 1.4% of the total surface area of South Africa. However, it is home to about 24% of the national population and is considered the economic hub of the country and the Southern African region, with its capital in Johannesburg.

In South Africa, the principles governing environmental protection are embedded in the Constitution of the country (Section 24, The Constitution of the Republic of South Africa 1996),[17] which empowers the government to take legislative action for the following:

- Prevent pollution and ecological degradation
- Promote conservation
- Secure ecologically sustainable development
- Use natural resources while promoting justifiable economic and social development.

Although there are a number of acts and regulations for different aspects of the environment, the key and overarching legislation for environmental management is the National Environmental Management Act No. 107 of 1998 (NEMA).[18] This Act is managed through two institutions: (1) the National Environmental Advisory Forum, which is a ministerial advisory body appointed by the Minister of Environmental Affairs and Tourism, and (2) the Committee for Environmental Co-ordination, which coordinates and integrates environmental functions of different state organs. The key pieces of national legislation that are used to enforce NEMA are specified in Schedule 3, Section 34 of NEMA, the key ones being Hazardous Substances Act No. 15 of 1989,[19] National Environmental Management: Air Quality Act No. 39 of 2004 (NEM:AQA),[20] National Environmental Management: Waste Act No. 59 of 2008 (NEM:WA)[21] and National Water Act No. 36 of 1998 (NWA).[22] These are normally referred to as the "Specific Environmental Management Acts" (SEMA). Both the NEMA and SEMA normally require the issuing of specific permits or licenses before certain classified activities are undertaken. However, there is currently no fully integrated permitting system – that is, a single environmental permit for all activities on site, thus there is a need to apply for separate permits under specific SEMAs. The NEMA Act is one of the most comprehensive pieces of legislation in SSA and addresses key aspects of sustainability, integrated pollution control, and energy and material consumption. It is mainly anchored on the "polluter pays principle", "prevention principle", "sustainable development principles", and the "public participation principle".

Waste management in South Africa is regulated by the National Environmental Management: Waste Act No. 59 of 2008 (NEM:WA). The objectives of the Act are to promote the zero-waste model, and where waste is unavoidable, to implement measures to reduce, reuse, recycle, and recover waste.

The enforcement of the NEMA and associated SEMA in each province is normally done through provincial or city bye-laws. In the City of Johannesburg, the key environmental bye-laws are as follows:

- Air Pollution Control By-laws 2021 of the City of Johannesburg Metropolitan Municipal[23]
- Waste Management By-laws 2021 of the City of Johannesburg Metropolitan Municipal[24]
- Treated Effluent By-law, 2017[25]

The City of Johannesburg's official waste management service provider is Pikitup Johannesburg (Pty) Limited, which was founded in January 2001 with the City of Johannesburg as its sole shareholder. Its main mandate, besides waste collection from residential and commercial households, is to minimize the waste stream going to landfills while extracting maximum value from the waste stream at all stages of collection and disposal.

4.6.2 Botswana

Botswana is a landlocked country located at the centre of Southern Africa with its capital in Gaborone. It shares its borders with South Africa, Namibia, Zambia, and Zimbabwe. It is a sparsely populated country with an estimated population of 2.3 million and covering an area of 581,730 square kilometres translating to an estimated population density of 4 people per square kilometre. Large parts of the country (approximately 70%) are comprised of the Kalahari Desert. It is considered to be one of Africa's stable democracies and has one of the fastest-growing economies in Africa. It is one of the world's largest producers of diamonds, which has helped the country transform into an upper middle-income economy.

In Botswana, there is no overarching environmental law, and there is no specific reference in the constitution relating to sustainable development. The main national pieces of legislation relating to environmental protection are as follows:

- Waste Management Act (Chapter 65:06)[26]
- Atmospheric Pollution (Prevention) Act (Chapter 65:03)[27]
- Environmental Impact Assessment Act, 2011 (No. 10 of 2011)[28]

- Radiation Protection Act (Chapter 24:03)[29]
- Water Act (Chapter 34:01)[30]

The Waste Management Act is managed by the Department of Sanitation and Waste Management, which is tasked with the responsibility to develop, manage, implement, and coordinate waste management and sanitation policies and plans in order to minimize pollution to the environment. The Act also makes reference to the Basel Convention in regulating the transboundary movement of hazardous wastes and their disposal. The Act also requires local authorities to come up with their local waste recycling plan, detailing the kind and quantity of waste that will be recycled and how this will be done. The Act does not fully address issues of sustainable development and the environmental legislation is based on single media control. The Environmental Impact Assessment Act is used as a preventative tool to assess the potential environmental effects of proposed developments and their mitigation, monitoring, and evaluation. Pollution control of water resources is partly covered in the Water Act but mainly managed through district or city bye-laws where standards for effluent discharges are specified (Central District Council (Public Sewer) Bye-laws, 2006).[31] Overall, there is no mechanism for integrated pollution control in the legislation, and issues of sustainability are not addressed. There is a need to develop an overarching environmental management act that will help in order to allow for the integration of the environmental legislation framework in the country. Currently, the country's commitment to sustainable development is reflected in its Vision 2036 document, Pillar No. 3, which speaks to Sustainable Environment (Vision 2036).[32]

4.6.3 Zimbabwe

Zimbabwe is a landlocked country in Southern Africa, sharing land borders with South Africa, Zambia, Botswana, and Mozambique, with its capital in Harare. It has an estimated population of 15.17 million and a land area of 390757 square kilometres translating to an estimated population density of 39 people per square kilometre. The backbone of the country's economy is agriculture and mining, with the country holding substantial reserves of about 40 minerals and coal. It also has one of the largest known coal-bed methane gas deposits in Africa.

The right to a clean environment and environmental protection is enshrined in Article 73 of the Constitution of Zimbabwe (Constitution of Zimbabwe Amendment (No. 1) Act, 2017).[33]

The constitution outlines the following measures to protect the environment:

- Prevent pollution and ecological degradation
- Promote conservation
- Secure ecologically sustainable development and use of natural resources while promoting economic and social development.

The overarching legislation for environmental management is the Environmental Management Act [Chapter20:27][34] which regulates all aspects of the environment such as sustainable resource management and environmental protection, pollution prevention, establishment of a National Environmental Plan, and other plans for the management and protection of the environment, the establishment of an Environmental Management Agency and an Environment Fund and the setting of environmental quality standards. The responsibility for managing the Act lies with the Environmental Management Agency (EMA) which oversees the implementation of the Act by different government departments. The Environmental Management Act is implemented through Statutory Instruments (SIs) based on section 140 of the Environmental Management Act (Chapter 20:27). The SIs establish the regulations for dealing with specific aspects of the environment. The main SIs that deal with waste are the following:

- Effluent and Solid Waste Disposal Regulations, SI No. 6 of 2007
- Plastic Bottles and Plastic Packaging Regulations, SI No. 98 of 2010
- Hazardous Substances, Pesticides and Toxic Substances Regulations, SI 268 of 2018

The Act is a general framework law largely based on the precautionary and polluter pays principles. The legislation does not currently fully address issues on the sustainable circular economy but focuses more on addressing integrated pollution control. A review of Zimbabwe's waste management policies geared towards achieving a circular economy revealed that the current regulations do not explicitly promote strategies towards transitioning to a circular economy (Nyakudya et al., 2022). They recommended that the country needs Zimbabwe requires a national strategy for a circular economy that collects and systematizes experiences, defines objectives and establishes clear goals, identifies and launches promising pilot projects, and provides information on the dimension of the existing benefits and barriers with a shift in mindset (Nyakudya et al., 2023).

4.7 EAST AFRICA

4.7.1 Kenya

Kenya is an East African nation with its capital in Nairobi and is the largest economy in East Africa. It has an estimated population of 55 million and occupies an area of 580,367 square kilometres translating to a population density of approximately 95 people per square kilometre.

The right to a clean and healthy environment is enshrined in the Constitution of Kenya,[35] however, the Environmental Management and Coordination Act (EMCA), 1999 (Chapter 387)[36] provides the framework for the appropriate legal and institutional arrangements for the management of the environment. This is the overarching environmental law and is administered through three main organs, namely, National Environmental Management Authority (NEMA), National Environment Council (NEC), and Provincial and District Environment Committees (PDEC). NEMA is the principal organ of the State in the implementation of all policies relating to the environment and manages and co-ordinates on all matters relating to the environment. NEC is responsible for policy formulation and setting of national goals for the protection of the environment. The PDEC is responsible for responsible for the proper management of the environment within their respective province or district and ensuring adherence to the EMCA. The Act also makes provision for the establishment of the Public Complaints Committee which provides the administrative mechanism for addressing environmental harm. Section 58 of the Act requires that an Environmental Impact Assessment Licence be obtained for any project that may or is likely to have a significant impact on the environment.

A number of Regulations and Acts have been established in line with the EMCA for specific environmental issues such as the Water Quality Regulations (Environmental Management and Coordination, (Water Quality) Regulations, 2006),[37] Environmental Management and Co-ordination (Waste Management) Regulations, 2006[38] and Sustainable Waste Management Act, 2021.[39] The Sustainable Waste Management Act makes provision for the implementation of sustainable circular economy in waste management. Kenya is one of the few African countries that have an Act that addresses the legal and institutional framework for sustainable management of waste. The Act promotes the implementation of the waste hierarchy (reduction, reuse, repair, refurbishment, recycling, recovery, and treatment for safe disposal) in order to achieve efficient use of resources and minimize pollution. It is based on the precautionary and polluter pays principles and seeks to implement the zero-waste model. It also promotes segregation and classification of waste at source as means of implementing the waste hierarchy and promotes circular economy practices for green growth.

4.7.2 Tanzania

Tanzania is an East African state and is one of Africa's fastest-growing economies. It has an estimated population of 61.7 million and occupies an area of 947,303 square kilometres translating to a population density of approximately 95 people per square kilometre. It is endowed with diverse natural resources and a youthful population, with approximately 45% of the population under the age of 15. Its capital is Dodoma and its commercial hub is Dar es Salaam which is the largest city in East Africa.

In Tanzania, the right to a clean and healthy environment and principles of sustainability are enshrined in the country's constitution (Article 9 of the Constitution of the United Republic of Tanzania (CAP. 2)).[40] The overarching environmental law is the Environmental Management Act, 2004.[41] The Act provides the legal and institutional framework for sustainable management of the environment and outlines principles for the management of the environment, impact and risk assessments, prevention and control of pollution, waste management, environmental quality standards, public participation, compliance and enforcement, and the basis for implementation of international instruments on the environment. In Tanzania, the Office of the Vice President is responsible for the environment, and the Act is administered through the National Environmental Advisory Committee, which is a ministerial advisory body. The Office of the Director of Environment is responsible for implementing the provisions of the Act, and the National Environment Management Council (NEMC) is tasked with undertaking enforcement, compliance, review, and monitoring of environmental impact assessment. The Act also establishes local government authorities, which are the principal executive agencies of environmental policies and regulations. It also promotes cleaner production technologies and techniques, fostering sustainable consumption of goods and services, and integrated pollution prevention and control. It also requires a mandatory environmental impact assessment (EIA) for all proposed activities that are likely to have significant adverse impacts on the environment. Provision is made in the Act for the formulation and approval of environmental standards and criteria that define regulatory limits for single media discharge into the environment. The National Environmental Standards Committee, established under the Tanzania Bureau of Standards Act, 1975,[42] is responsible for developing and reviewing the environmental discharge standards. The environmental law in Tanzania is largely modelled on the precautionary and polluter pays principles. Sustainable circular economy principles are not currently addressed. Issues on sustainable development are mainly addressed in the National Environmental Policy, 1997,[43] which provides a set of principles and objectives for an integrated and multi-sectoral approach addressing the totality of the environment (Pallangyo, 2007).

A number of regulations have been developed for the enforcement of the Environmental Management Act, 2004, such as the Environmental (Solid Waste Management) Regulations, 2009,[44] Environmental Management (Control and Management of Electrical and Electronic Equipment Waste) Regulations, 2021,[45] Environmental Impact Assessment and Audit Regulations, 2005,[46] Environmental Management (Prohibition of Plastic Carrier Bags and Plastic Bottle Cap Seals) Regulations, 2022,[47] Environmental Management (Hazardous Waste Control and Management) Regulations, 2021,[48] and Environmental Management (Air Quality Standards) Regulations, 2007.[49] The regulations on solid waste management are mainly based on segregation at source, integrated solid waste management (including recovery and recycling of wastes), and waste minimization at source through the implementation of cleaner production principles.

4.8 CENTRAL AFRICA

4.8.1 Rwanda

Rwanda is a landlocked country in the Great Lakes Region with its capital in Kigali. It is a relatively small country occupying an area of 26,338 square kilometres and an estimated population of 13.8 million people, which translates to a population density of 524 people per square kilometre. It is a densely populated country and has the highest population density in Africa. The population, however, is unevenly distributed, with an estimated 95% of the population living in rural areas. The most densely populated region is Ruhondo in Ruhengeri.

The principal environmental law in Rwanda is the Law N°48/2018 of 13/08/2018 on Environment,[50] which is anchored in the Constitution of the Republic of Rwanda (Article 22: Right to a clean environment and Article 53: Protection of the environment).[51] The law describes the modalities for protecting, conserving, and promoting the environment. The law is comprehensive and covers all aspects of the environment, such as the protection of the natural environment (soil, water bodies, wetlands, and the atmosphere); regulation of effluent discharges and solid waste to the environment; financing of environmental issues and penalties for non-compliance; and energy use. It also details the roles of the state, decentralized entities, and local communities with regard to the protection, conservation, and promotion of environmental issues. Environmental impact assessments and environmental audits are also mandatory for gazetted projects as specified by the Order of the Minister in charge of the environment.

The law is largely based on the precautionary and polluter pays principles. It, however, does make reference to the principle of environmental sustainability, although this is not fully explored. The agency responsible for environmental management is the Rwanda Environment Management Authority (REMA)[52] which is under the Ministry in charge of the environment. REMA is mainly responsible for implementing government environmental policy, establishing measures to prevent and mitigate climate change, and pollution control, ensuring environmental regulations for sustainable development are complied with, and coordinating, monitoring, and evaluating environmental impact assessments and environmental audits. Rwanda's National Environmental Policy, 2003[53] sets out overall and specific objectives as well as fundamental principles for improved management of the environment, both at the central and local levels. Overall, the country has comprehensive environmental laws and management structures that can enable it to achieve a sustainable circular economy.

4.8.2 Cameroon

Cameroon is a central African country with its capital in Yaoundé. It occupies a land area of 475,442 square kilometres with an estimated population of 29.3 million. This translates to a population density of approximately 62 people per square kilometre. It is a lower middle-income country endowed with a number of natural resources such as oil, gas, and mineral resources.

The main environmental management law in Cameroon is the Law 96/12 of August 1996 Relating to Environmental Management[54] (hereafter referred to as the "1996 Law"). The "1996 Law" lays down the legal framework for environmental management in Cameroon. It details the coordination and financing of environmental policies; the establishment of the National Environmental Plan, which is amended every 5 years; mandatory environmental impact assessments for any project or development that may endanger the environment; protection of the environmental bodies (atmosphere, soil, waters, and floodplains) and human settlements by regulating environmental discharges; waste and hazardous waste management; incentives and penalties for noncompliance; and implementation and monitoring of environmental programmes. The 1996 Law is mainly a framework law and does not go into details as far as how the principles of environmental protection will be enforced or implemented. The "1996 Law" is largely based on the precautionary principle and the polluter pays principle. Although reference is made to the public participation principle, it is not fully developed in the law. While the principle of sustainable development is defined in the law, it is not fully explored anywhere within the "1996 Law". The law also focuses on single media control, and integrated pollution control is not addressed. Thus, the law currently falls short of addressing the current environmental challenges of modern-day society.

The government agency responsible for environmental management in Cameroon is the Ministry of the Environment, Protection of Nature and Sustainable Development (MINEPDED). MINEPDED has published in a single document most of the laws and regulations relating to environmental management in Cameroon. The objective of the document was to integrate the principles of sustainable development into national policymaking, achieve sustainable management of natural resources, and fight against pollution and nuisances.[55] A comprehensive analysis of Cameroon's environmental laws and policies is also covered by Ruppel and Yogo (2018).

4.9 SUMMARY FINDINGS

While all SSA countries have legislation to address environmental management, they differ in terms of content and scope. For a majority of the countries, the right to a clean environment and the right to protect the environment is enshrined in the national constitution. The dominant principles upon which a majority of the environmental management laws are based are the "polluter pays principle" and the "precautionary principle". Only a few countries have legislation based on "sustainable development principles", although this is defined in most of the legislation, it is never fully addressed in the environmental laws. South Africa and Kenya were the only countries considered in this chapter that have legislation to address sustainable development and the circular economy. For some of the countries such as Tanzania, Rwanda, and Botswana policies have been developed to transition the countries towards a circular sustainable economy, however, laws and regulations in this regard have not been developed as yet. A majority of the legislation is still designed based on single media control, and integrated pollution control is never addressed. South Africa has the most comprehensive set of environmental laws in SSA.

The current major challenge in SSA is the enforcement of the existing legal framework. It has been noted that enforcement is a challenge in most SSA countries as a result of a lack of human capacity, working tools, analytical facilities, lack of political will, and financial constraints (Pallangyo, 2007; Debrah et al., 2022; Sabit, 2022). These challenges can be partly addressed by the coordination of environmental activities at a level such as the African Union to enable countries to share resources and expertise.

NOTES

1 https://www.aceaafrica.org/about-acea
2 https://lagosstate.gov.ng/
3 https://nigerian-constitution.com/
4 https://www.placng.org/lawsofnigeria/laws/E12.pdf
5 https://www.placng.org/lawsofnigeria/laws/F10.pdf
6 https://lawsofnigeria.placng.org/laws/nesrea.pdf
7 https://faolex.fao.org/docs/pdf/nig188382.pdf
8 https://lasepa.lagosstate.gov.ng/
9 https://lawsofnigeria.placng.org/laws/nesrea.pdf
10 http://nsagm.weebly.com/uploads/1/2/0/3/12030125/national_environment_management_act_1994.pdf

11 https://nea.gm/overview/
12 https://nea.gm/overview/
13 https://www.basel.int/Portals/4/download.aspx?d=UNEP-CHW-NATLEG-NOTIF-Gambia01-ACT2003.English.doc
14 https://faolex.fao.org/docs/pdf/gam173305.pdf
15 https://constitutionnet.org/sites/default/files/Ghana%20Constitution.pdf
16 https://www.fao.org/faolex/results/details/en/c/LEX-FAOC013234/
17 https://www.gov.za/documents/constitution/chapter-2-bill-rights
18 https://www.gov.za/documents/national-environmental-management-act
19 https://www.gov.za/documents/hazardous-substances-act-16-apr-2015-1120
20 https://www.gov.za/documents/national-environment-management-air-quality-act
21 https://www.gov.za/documents/national-environmental-management-waste-act
22 https://www.gov.za/documents/national-water-act
23 https://www.joburg.org.za/documents_/Documents/By-Laws/2022/City%20of%20Johannesburg%20Air%20Pollution%20Control%20By-laws%2013%20October%202021.pdf
24 https://www.joburg.org.za/documents_/Documents/By-Laws/waste%20management%20by-law.pdf
25 https://openbylaws.org.za/akn/za-jhb/act/by-law/2017/365/eng@2017-04-26
26 https://www.gov.bw/sites/default/files/2020–03/Waste%20Management%20Act.pdf
27 https://faolex.fao.org/docs/pdf/bot42110.pdf
28 https://www.gov.bw/sites/default/files/2020–02/Environmental%20Assessment%20Act%202011.pdf
29 https://faolex.fao.org/docs/pdf/bot91332.pdf
30 https://faolex.fao.org/docs/pdf/bot42103.pdf
31 https://botswanalaws.com/StatutesActpdf/2006Subsidiarypdf/CENTRAL%20DISTRICT%20COUNCIL%20(PUBLIC%20SEWER)%20BYE-LAWS,%20S.%20I.%20NO.%2032%20OF%202006.pdf
32 https://www.statsbots.org.bw/sites/default/files/documents/Vision%202036.pdf
33 https://faolex.fao.org/docs/pdf/zim127325.pdf
34 https://zw.chm-cbd.net/sites/zw/files/inline-files/EMA%20ACT.pdf
35 https://faolex.fao.org/docs/pdf/ken127322.pdf
36 https://eregulations.invest.go.ke/media/emca_1.pdf
37 http://kenyalaw.org/kl/fileadmin/pdfdownloads/LegalNotices/2006/LN120_2006.pdf
38 https://www.nema.go.ke/images/Docs/Regulations/Waste%20Management%20Regulations-1.pdf
39 http://www.parliament.go.ke/sites/default/files/2022-05/The%20Sustainable%20Waste%20Management%20Bill%20%28%20National%20Assembly%20Bills%20No.%2022%29%202021.pdf
40 https://rsf.org/sites/default/files/constitution.pdf
41 https://www.nemc.or.tz/uploads/publications/sw-1576228517-ema%20act2004.pdf
42 https://tanzanialaws.com/s/345-standards-act
43 https://www.nemc.or.tz/uploads/publications/sw-1576228807-NEP%201997.pdf
44 https://www.parliament.go.tz/uploads/documents/1476206361-GN%20190-THE%20ENVIRONMENTAL%20MANAGEMENT%20(SOLID%20WASTE%20MANAGEMENT)%20(AMENDMENT),%202016.pdf
45 https://www.nemc.or.tz/uploads/publications/sw-1645446706-e_WASTES%20REGULATIONS_2021.pdf
46 https://www.nemc.or.tz/uploads/publications/sw-1576227418-Revised%20-%20EIA%20and%20Audit%20Regulations%202018%20-%20NEW.pdf
47 https://faolex.fao.org/docs/pdf/tan215725.pdf
48 https://www.nemc.or.tz/uploads/publications/sw-1645446901-HAZARDOUS%20WASTES_REGULATIONS%202021.pdf

49 https://www.nemc.or.tz/uploads/publications/sw-1645446559-Air_Quality_Standards_Regulations_2007.pdf
50 https://climate-laws.org/document/law-ndeg48-2018-on-the-environment_5d99
51 https://faolex.fao.org/docs/pdf/rwa128551.pdf
52 https://www.rema.gov.rw/home
53 https://rema.gov.rw/fileadmin/templates/Documents/rema_doc/Policies/Rwanda%20Environmental%20Policy_English.pdf
54 https://www.snh.cm/images/reglementation/EN/Law%20Environment.pdf
55 https://www.gfbcam.com/wp-content/uploads/2021/06/RECEUILS_DES_TEXTES_JURIDIQUES_MINEPDED_ANGLAIS-1.pdf

REFERENCES

Akuffo, S. B. (1998). *Approaches to pollution control, pollution control in a developing economy: A study of the situation in Ghana* (Chapter 2, pp. 31–43). Ghana Universities Press.

Coelho, P. M., Corona, B., ten Klooster, R., & Worrell, E. (2020). Sustainability of reusable packaging: Current situation and trends. *Resources, Conservation & Recycling: X, 6*, 100037. ISSN 2590-289X. https://doi.org/10.1016/j.rcrx.2020.100037

Commoner, B. (1988). Economic growth and environmental quality: How to have both. *Environment Annual Editions, 87–88*, Article 6, 47–55.

Debrah, J. K., Teye, G. K., & Dinis, M. A. P. (2022). Barriers and challenges to waste management hindering the circular economy in sub-Saharan Africa. *Urban Science, 6*, 57. https://doi.org/10.3390/urbansci6030057

EC. (1996). *Council directive 96/61/EC concerning integrated pollution prevention and control, official journal of the European communities, L257(10/10/96)*. EC.

Eurostat. (2019). *Municipal waste and packaging waste statistics*. Eurostat.

Korhonen, J., Honkasalo, A., & Seppälä, J. (2018). Circular economy: The concept and its limitations. *Ecological Economics, 143*, 37–46. https://doi.org/10.1016/j.ecolecon.2017.06.041

Nicholas, M. J., Clift, R., Azapagic, A., Walker, F. C., & Porter, D. E. (2000). Determination of "best available techniques" for integrated pollution prevention and control: A life cycle approach. *Process Safety and Environmental Protection, 78*(B3), 193–203.

Mensah, A. E. *et al.* (2020). 'Does Environmental Laws and Policies Work? A Review Of Ghana's Case', *Research in Ecology, 2*(3), 32–41. doi: 10.30564/re.v2i3.2216

Nyakudya, P., Madyira, D., & Madushele, N. (2022, July). Seeking Circularity: A Review of Zimbabwe Waste Management Policies Towards a Circular Economy. In *5th European International Conference on Industrial Engineering and Operations Management*, IEOM Society International. https://doi.org/10.46254/EU05.20220407

Pallangyo, D. M. (2007). Environmental law in Tanzania; how far have we gone? *Law, Environment and Development Journal, 3*(1), 26. http://www.lead-journal.org/content/07026.pdf

Pollution Prevention Act (PPA, 1990), 42 U.S.C 13101 and 13102, S/S et seq.

Ruppel, O. C., & Yogo, E. D. K. (Eds.). (2018). *Environmental law and policy in Cameroon—towards making Africa the tree of life | Droit et politique de l'environnement au Cameroun—Afin de faire de l'Afrique l'arbre de vie* (1st ed.). Nomos Verlagsgesellschaft mbH. http://www.jstor.org/stable/j.ctv941sr6

Sabit, K. (2022, May 9). *Environmental laws in Nigeria: An overview*. SSRN. https://ssrn.com/abstract=4104055 or http://dx.doi.org/10.2139/ssrn.4104055

UNCED. (1992, June). *Agenda 21*. United Nations Conference on Environment and Development.

United Nations Environment Programme. (2018). *Africa waste management outlook—summary for decision-makers*. Retrieved April 17, 2022, from https://wedocs.unep.org/20.500.11822/25515

5 Current Municipal Solid Waste Management in Africa

Thandiwe Sithole

5.1 INTRODUCTION

Due to the lack of waste disposal sites, MSW has become a serious environmental and health issue. More waste is generated daily due to the rapid relocation of the rural population to new regions and countries in order to find work and better living conditions. This greatly influences waste generation and management. Three categories of waste greatly contribute to the generation of municipal solid waste. These include, but are not limited to, domestic waste, which consists of waste from restaurants, households, markets, and other commercial spaces. Industrial waste comprises waste from private industries but excludes waste that requires special handling. Institutional waste covers waste from hospitals, public institutions, schools, establishments, and recreational areas. As such, mismanagement of MSW not only has a negative environmental impact but also causes risks to public health and raises various socioeconomic issues that are worth discussing. As a result, it is crucial to urgently improve the handling of waste collection, isolation, and safe disposal.

Every single one of us creates a sizable quantity of solid waste every day. Any material that is unwanted or unusable is considered solid waste, and this includes electronics, paper, plastic, glass, textiles, and yard waste (Baawain et al., 2017). A large number of this waste is produced by homes and businesses, though some waste material is also created by mineral extraction, agricultural production, and energy production (Ayeleru, 2016). The concept of "solid waste" (SW) refers to all items that have been deemed useless and that the original generators do not want to maintain anymore (Christensen, 2011). The alternative term is "trash." SW mainly comes from urban residential and business areas. Depending on the source, SW can be divided into eight different kinds. These classes of solid waste are as follows: residential, industrial, heavy industry, agricultural, municipal services, commercial, institutional, construction, and demolition (Ayeleru, 2016).

According to Ayeleru (2016), residential garbage comes primarily from small- and big-family residential houses. Residential garbage is made up of ineffective materials such as cardboard, plastics, textiles, leather, yard waste, wood, glass, metals, ashes, electronics, food scraps, and papers. This includes all waste produced at home or where we live. Industrial waste includes household wastes, packaging, food wastes, construction and demolition wastes, hazardous wastes, ashes, special

 DOI: 10.1201/9781003377856-5

wastes, chemical plants, and so on (Woodard, 2001). Institutional squandering is derived from educational institutions, hospitals, jails, and government parastatals. Cardboard, plastics, paper, food waste, metals, specific wastes, glass, and so on are all included (Ayeleru, 2016).

Construction and demolition debris is generated at building construction sites, road construction sites, and renovation sites (Mália et al., 2013). It contains sluggish materials like rubble and bulky building waste, as well as timber, steel, concrete, and lumber. Municipal services trash includes street sweepings, tree trimmings, parks and recreation centres, wastewater treatment facilities, beaches, and so on (Zhou et al., 2019). Waste from the processing or manufacturing industries includes oil refineries, chemical processing garbage, material waste, pesticides of different plants, power plants, extraction and processing, and so on. Agricultural trash is generated by gardens, animal husbandry, damaged food waste, agricultural garbage, farms, and so on (Ayeleru, 2016).

5.2 MUNICIPAL SOLID WASTE

Any material or object derived from domestic hazardous trash, community activities, or commercial or institutional operations is considered municipal solid waste (MSW), but radioactive and potentially dangerous waste are not included (Damanhuri et al., 2014). According to Fauziah and Agamuthu (2012), any waste that is under the jurisdiction of the municipal government and for which appropriate collection and disposal plans have been developed is referred to as MSW. Food waste, packaging, street sweeper runoff, yard waste, recyclables, by-products of industrial production, and other waste are all included in MSW. Due to a number of factors, the generation of MSW varies by location (Achankeng, 2003).

The high standard of living, the increased rural-to-urban migration, the advancement of industry and the economy, as well as climatic and geographic changes, are some of these factors. Municipalities are in control of disposing of waste in Africa (Aliu et al., 2014). Due to factors like limited municipal funding, staff shortages, an ageing vehicle fleet, long travel distances required to provide services in rural areas, widespread illegal dumping, rapid urbanization, and the rise of informal neighbourhoods and backyard dwellings, providing waste services is particularly challenging. Transporting and collecting waste is a crucial component of providing waste management services (Loukil & Rouached, 2020).

Governments must function effectively due to the high cost of waste collection and the scarce financial resources available in many communities (Korfmacher, 1997). Efficiency in internal operations refers to maximizing outputs, like waste removal services, with the resources available, like time, trucks, staff, and financial resources. It should be noted that even though the country's many municipalities offer an efficient trash collection service at service agreement levels, it is still possible that they aren't meeting their obligations (Loukil & Rouached, 2020).

MSW, which was produced in Africa in 2012 in quantities of 125 million tonnes, is anticipated to increase in the coming years. That any reduction in garbage creation in other regions would be dwarfed by Africa. This is due to insufficient garbage collection in the majority of African countries. MSW collection rates are only 55% on average (Godfrey et al., 2019). Figure 5.1 shows the projected total amount of MSW generated in Africa.

More than 90% of the trash in Africa is disposed of in illegal dumps and landfills (Rasmeni & Madyira, 2019), often in conjunction with open burning. All of the 19 largest dumpsites in the world are located in sub-Saharan Africa. The majority of the organic waste, which makes up 57% of MSW created in Africa, is currently discarded but may have significant socioeconomic potential for nations. In contrast, 13% of MSW generated in Africa is plastic. Africa is experiencing an increase in recycling, driven more by socioeconomic needs, unemployment, and poverty than by public or private sector planning. Only 4% of MSW is recycled at the moment, although 70–80% of MSW produced in Africa is thought to be recyclable. Informal waste pickers actively participate in recovering valuable resources from trash at little to no cost to local governments and private companies (Godfrey et al., 2019).

Over the coming century, sub-Saharan Africa, in particular, will experience significant social and economic change as a result of population growth, urbanization, and changing consumer spending patterns (Godfrey et al., 2019). We can conclude this from the results in Figure 5.2. Major increases in trash production are likely to follow, further taxing the already overworked infrastructure and garbage services in the public and private sectors. Figure 5.3 shows a comparison of MSW composition for sub-Saharan Africa and globally.

How waste is gathered and dumped seems to have a direct impact on its composition (Okot-Okumu, 2012). Table 5.1 provides a general breakdown of the amount of carbon-based (organic), plastic, glass, metal, paper, and other materials in MSW (Ayeleru, 2016). Sub-Saharan Africa produces 57% of the world's MSW from organic waste, which is significantly more than the global average (relative to the other waste streams), according to the World Bank. Furthermore, sub-Saharan Africa has a higher plastic content, which is 13% of MSW, compared to the rest of the world (*Africa Waste Management Outlook 3 state of solid waste management in Africa, n.d.).*

TABLE 5.1
Types and Sources of Waste

Type	Sources
Plastic	Bags, cups, lids, bottles, and other packaging.
Metal	Railings, bicycles, non-hazardous aerosol cans, cans, foil, and tins are examples of white goods.
Glass	Glassware in pieces, bulbs, coloured glass, and bottles.
Paper	Paper waste includes things such as magazines, cardboard, bags, boxes, phone books, wrapping paper, newspapers, shredded paper, and paper cups for drinking.
Organic	Remaining food, yard waste (including grass, leaves, and brush), wood, and manufacturing waste.
Others	Leather, rubber, electronics waste, ash, laminates, textiles, appliances, and other materials that are inert.

Source: Bhada-Tata and Hoornweg (2016)

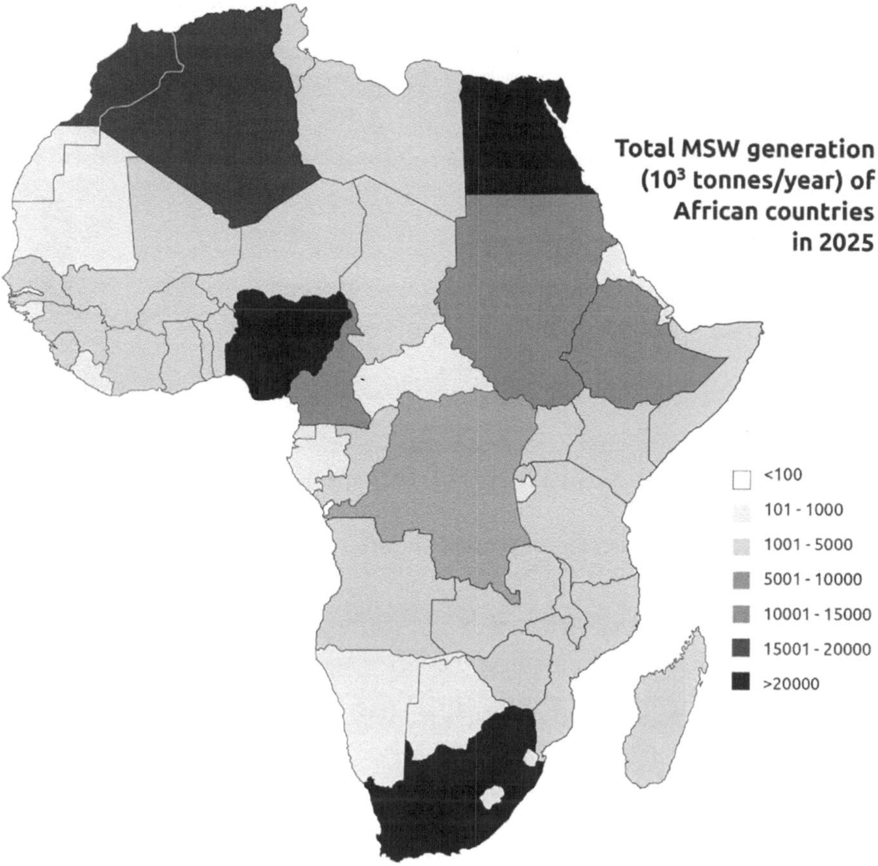

FIGURE 5.1 Africa map showing MSW in all countries in millions of tonnes.

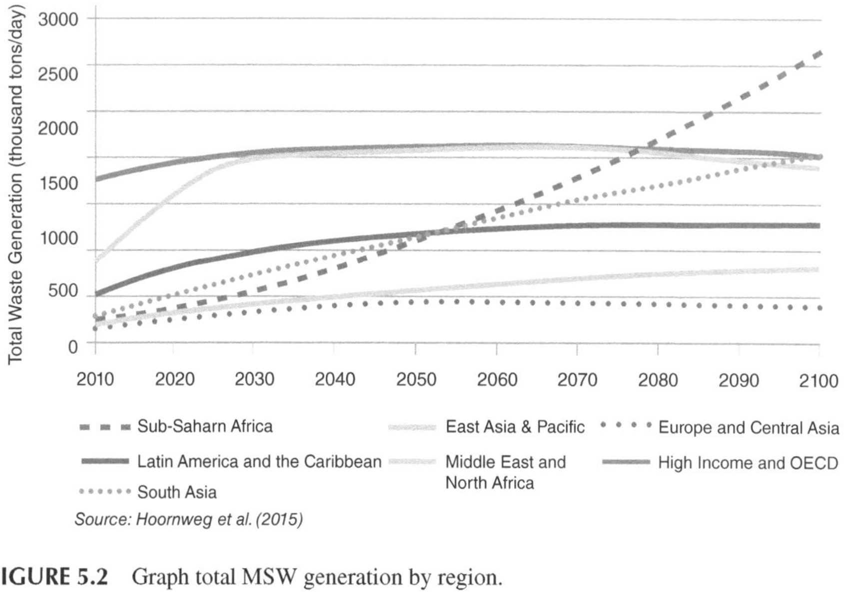

FIGURE 5.2 Graph total MSW generation by region.

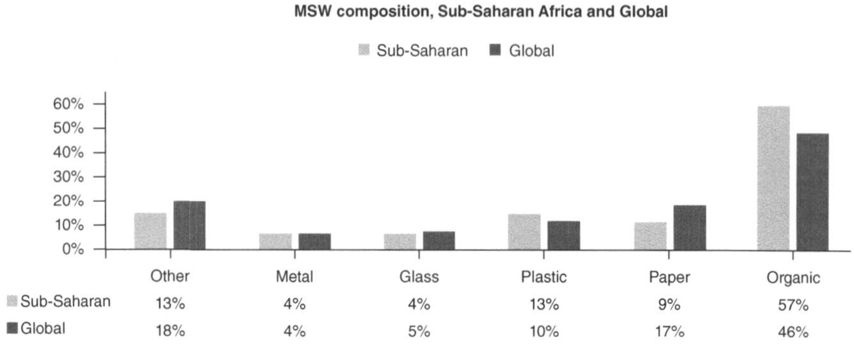

FIGURE 5.3 MSW composition, sub-Saharan Africa.

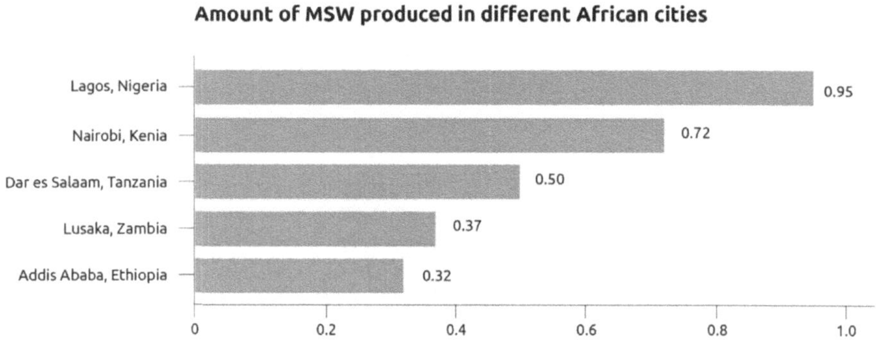

FIGURE 5.4 Quantity of MSW generated in various African cities.

Every city has a different mix of MSWs, which differ based on things like consumer attitudes, socioeconomic status, and culture. According to data collected for twelve African cities (Table 5.2), the average percentage of organic waste in all MSW is over 60%, with significant variation between cities. Organic waste accounts for a significant portion of garbage generated in low- and middle-income areas, owing primarily to fresh food preparation, whereas waste in high-income cities is more diverse, with relatively larger shares of paper and packaging, including plastics (Bhada-Tata & Hoornweg, 2016).

Despite constituting less than 10% on average of MSW in cities, plastic waste is a significant contaminant in Africa. MSW produced in Africa has high moisture content because of the high proportion of organic waste present (Damanhuri et al., 2014), which has an immediate effect on waste management, the potential environmental effects of the waste when dumped in landfills, and the suitability of alternative waste treatment technologies adopted in Africa. The composition of MSW generated in selected African countries as shown in Figure 5.4.

TABLE 5.2
MSW Composition for Selected African Cities

City and Reference	Organic%	Paper/ cardboard%	Plastic%	Glass%	Metal%	Others%
Kampala, Uganda Bello et al. (2016)	77.2	8.3	9.5	1.3	0.3	3.4
Dar es Salaam, Tanzania Bello et al. (2016)	71.0	9.0	9.0	4.0	3.0	4.0
Ibadan, Nigeria Adeyi and Adeyemi (in press)	69.6	7.67	4.47	2.00	1.65	14.6
Accra, Ghana Oteng-Ababio et al. (2013)	65.0	6.0	3.5	3.0	2.5	20.0
Moshi, Tanzania Bello et al. (2016)	65.0	9.0	9.0	3.0	2.0	12.0
Sousse, Tunisia UN-Habitat (2010)	65.0	9.0	9.0	3.0	2.0	11.0
Nairobi, Kenya UN-Habitat (2010)	65.0	6.0	12.0	2.0	1.0	15.0
Lagos, Nigeria Adeyi and Adeyemi (in press)	62.6	10.7	4.2	2.5	2.2	19.7
Abuja, Nigeria Imam et al. (2008)	56.3	11.4	10.2	3.9	5.2	N/A
Cairo, Egypt UN-Habitat (2010)	55.0	18.0	8.0	3.0	4.0	12.0
Tshwane, South Africa Komen et al. (2016)	53.8	11.5	9.5	6.7	1.8	16.7
Windhoek, Namibia Harts and Smith (2008)	48.0	15.0	11.0	14.0	4.0	8.0
Average	**62.8**	**10.1**	**8.3**	**4.0**	**2.5**	**12.4**

(*abbreviation:* N/A, not available).

5.3 MANAGEMENT PRACTICES

Africa's current waste management practices influence the economy, society, and environment (Gumbo & Simelane, 2015). African consumers now use more plastic, which, when combined with insufficient MSW collection systems, puts the continent at risk of increasing marine plastic litter (Loukil & Rouached, 2020). This is due to changing consumer behaviour. Growing pollution would severely affect coastal economies in Africa if no action were taken to stop the flow of plastic and other garbage into the ocean (Horejs, 2020). Uncontrolled trash disposal is common in urban areas, raising the risk of disease, flooding, and environmental degradation.

Anaerobic decomposition, which can harm ground and surface water, is also produced when organic waste is disposed of, contributing to climate change (Gumbo & Simelane, 2015).

Municipal solid waste management (MSWM), according to Achankeng (2003), is a significant environmental issue in many African cities. The majority of MSWM services in Africa are offered by the unorganized sector, which includes waste collectors, recycling facilities, and waste pickers (Gumbo & Simelane, 2015). Although the unofficial sector plays a significant role in MSWM, it is frequently unorganized and unregulated, which can have detrimental implications for the environment and public health (Fuller et al., 2022).

Gumbo and Simelane (2015) claim that open burning or combustion of waste results in significant air pollution, which has a detrimental effect on human health. The 130 or more people who died in recent African landfill collapses were made up primarily of women (Fauziah & Agamuthu, 2012). In Africa, gender inequalities in waste management have largely gone unnoticed. End-of-life products, such as electronic waste, which typically come from wealthy nations, are now dumped in Africa. Usually, the infrastructure for recycling and treatment of waste is insufficient to safely handle these waste streams, which has an immediate negative impact on both human health and the environment (Nithya et al., 2021). Figure 5.5 shows the African demographics of mismanaged plastic in Africa by 2010 and estimates in 2025.

5.3.1 Waste Collections

Waste collection and management will always be considered one of the most important services carried out by the municipal. However, in developed countries with all the resources, this is a common practice. In Brazil, solid waste management is regulated by the National Solid Waste Policy (Oh et al., 2018), which was effective in 2010 (Brazil, 2010). Municipal governments mount voluntary drop-off centres for the population and execute selective collection. Also, waste pickers partake in a system of door-to-door selective collection (CEMPRE, 2019). If municipal governments cannot properly manage waste, they can hardly provide any other service to their citizens (Hoornweg & Bhada-Tata, 2012).

In developing countries with low income, regulations are often limited; available resources are rare, and other inefficient methods for final waste disposal are regularly used. According to Mochache et al. (2020), in Emburu 37% of the households discard their solid waste in open places, 32% burn it, and 24% recycle it. Only 8% of households salvage their solid waste. In the small town Asella in Ethiopia, most of the residents practice improper solid waste management. The lack of sufficient data about solid waste management and the restrictions on collecting the waste could have contributed to improper solid waste disposal (Lema et al., 2019).

The most popular ways to eliminate waste in Africa involve using open dump sites (Carlos-Alberola et al., 2021). In the case of South Africa, garbage bags and bins are a common means of household waste collection done on a weekly basis by private waste contractors or taken to communal dumpsites or central collection points. Motorized collection vehicles with compactor vehicles are normally used for collection, while communal collection is common in areas without a formal refuse removal

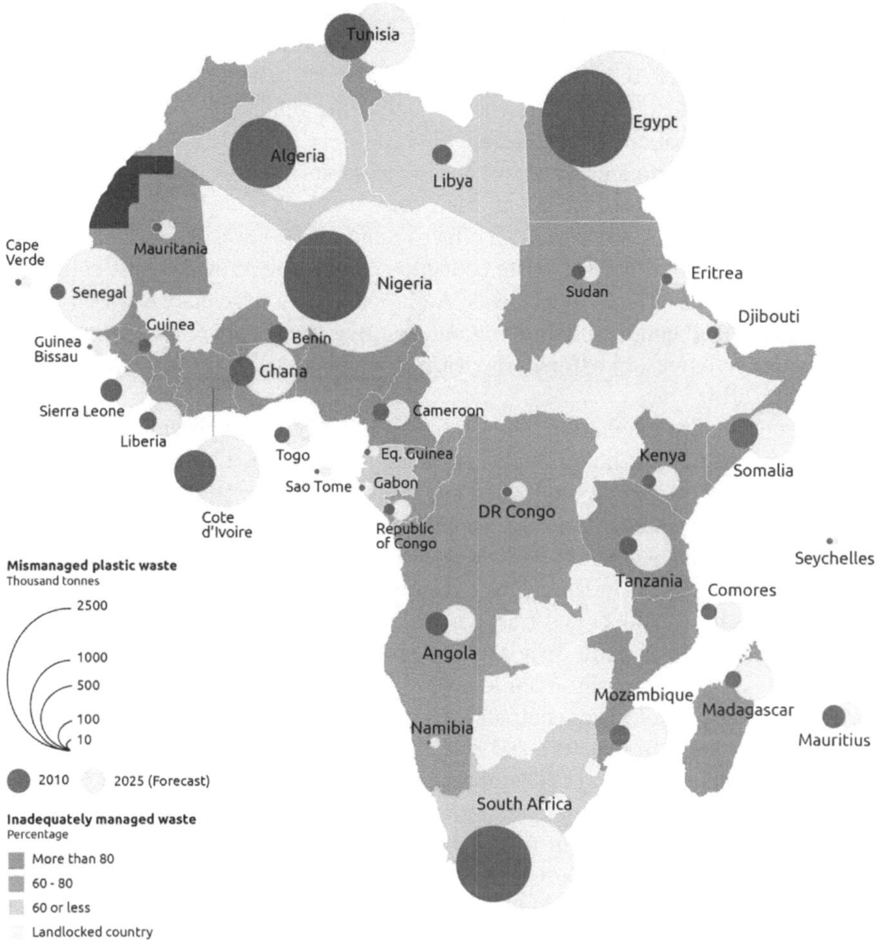

FIGURE 5.5 Africa region showing mismanaged plastic in Africa by 2010 and estimated in 2025.

service (Adeleke et al., 2021). In the same sense, in rural areas, residents burn and bury their waste materials because there are no municipal waste collection services.

Lately, residents have opted to recycle waste materials such as steel, cans, paper, and plastic, to mention a few. A decline was evident in the percentage of household removal rates from 2001 to 2016 in major metropolitan cities of South Africa, due to the gradual incline in the population of informal settlements with narrow access to collection services. Urban cities have about 90% more access compared to rural households with about 47% waste services. About 59% of household waste is collected by the local authority, 34% is disposed of in communal dumpsites, and 2% is collected from central collection points (DEA, 2018 #1).

As can be seen, waste collection requires a collective effort from the whole population in order to achieve a highly effective selective collection and recycling. Waste transportation and collection are considered crucial parts of waste disposal and are driven by concerns for the environment and general welfare (Aliu et al., 2014). Waste transportation refers to the act of moving collected waste to its final location, whereas waste collection refers to the use of rubbish-collecting vehicles to load garbage from waste collection locations (Korfmacher, 1997).

Due to inadequate financial support and other factors like lack of political consent, lack of priority, inadequate technical proficiency, and lack of dedication, municipalities frequently struggle to provide an effective and efficient waste collection service (Korfmacher, 1997). Effective waste collection can be determined by the calibre or standard of services offered to residents. At the same time, efficiency relates to internal operations and entails maximizing outputs, like waste collection services, with the available resources, like time, vehicles, personnel, financial resources, and others (Aliu et al., 2014).

It is important to emphasize that towns can be effective at providing waste collection services, but that does not always mean that they are efficient. Since internal effectiveness affects garbage collection costs, ineffective operational behaviour may lead to excessive spending (Loukil & Rouached, 2020). This is crucial because waste collection, which accounts for 60–80% of overall costs, is frequently regarded as the most expensive aspect of waste management (Loukil & Rouached, 2020).

It is necessary to collect and transport the waste produced for recycling, beneficiation, treatment, or landfill disposal. The provision of trash collection services to communities is the responsibility of local governments. Systems for managing waste involve a wide range of inputs, outputs, and actions. Depending on the type of waste produced, and the infrastructure and systems in place in the municipal, waste goes through a variety of processes (Dlamini et al., 2019). Figure 5.6 shows the structure of the collection and treatment operation of solid waste.

Even though the majority of funds allocated for the management of solid waste in developing nations go towards garbage collection (Aliu et al., 2014), the amount of MSW collected in Africa in 2012 was only estimated to be 55% of that produced, or 68 million tonnes (Loukil & Rouached, 2020). The normal municipal solid waste collection rate in sub-Saharan Africa was estimated to be only 44%. African countries' collection rates range from as low as 18% to more than 80%. By 2025, the total collection rate for the African region is anticipated to reach 69% (Aliu et al., 2014). Despite this rise in collection rate, the amount of uncollected MSW is not predicted to decrease, given the likely growth in MSW output. As a result, it is anticipated that the challenge of collecting MSW in Africa will continue to increase through the year 2025, posing a threat to human health (Sulemana et al., 2018). Figure 5.7 shows the waste collection and transportation process.

- **Process overview**

  Stage 1: The waste collection truck leaves the depot or garage for the collection location.

  Stage 2: When the vehicle arrives at the collection station, it stops and loads and unloads garbage from bins into the van. This is done along a predetermined path.

  Stage 3: When the vehicle is fully loaded, it is driven to a treatment or disposal facility of waste.

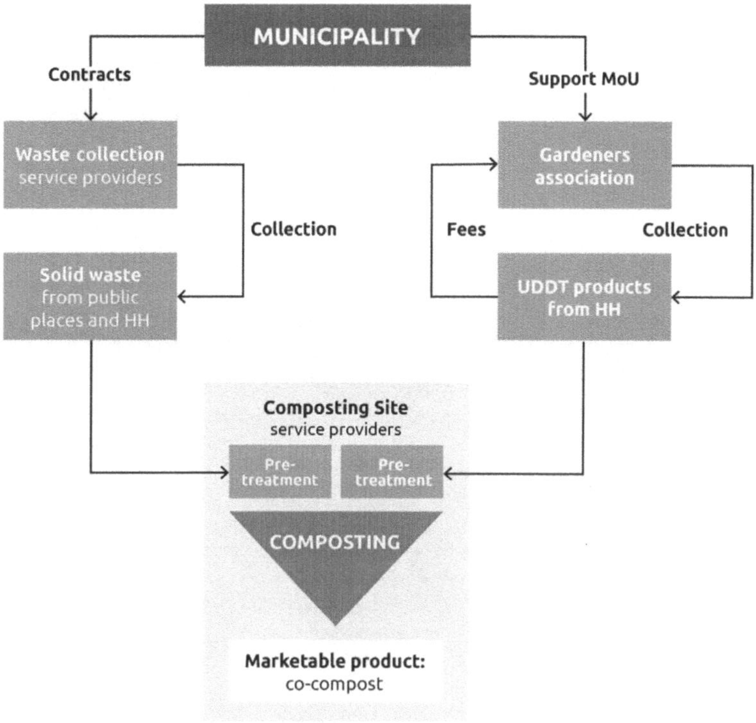

FIGURE 5.6 Structure of the collection and treatment operation.

Stage 4: After unloading at a facility for recovery, beneficiation, treatment, or disposal, the car travels back to the collection route to restart the procedure (Sulemana et al., 2018).

- **Issues that prevent effective and efficient waste collection services from being provided.**

The most frequent issues facing solid waste collection systems are financial constraints, fleet ageing, inefficient vehicle routing, poor road networks, personnel issues, and organizational challenges (Korfmacher, 1997). Waste collection is a problem in rural areas, rapidly growing urban areas, and squatter settlements. Since good route planning is necessary to lower operating costs, reduce the municipal's carbon footprint, and increase time efficiency, waste collection vehicle drivers cannot be trusted to choose routes (Sulemana et al., 2018). According to Loukil and Rouached (2020), an effective collection schedule for garbage is necessary in order to reduce costs and protect the environment.

Garbage management is often allocated a minor portion of the municipal budget, with revenue earned through waste charges. However, waste management activities cannot be carried out if tariffs are set insufficiently and users do not pay for trash services (Okot-Okumu, 2012). It costs money to, for instance, hire skilled and unskilled labour, buy vehicles

FIGURE 5.7 Waste collection and transportation process.

and equipment for garbage collection, maintain and refuel collection vehicles, and maintain waste management infrastructure (Sulemana et al., 2018). Waste collection is a fuel-intensive activity in garbage management systems that significantly raises the overall cost of municipal waste management.

5.3.2 Waste Processing

Waste processing is the treatment of solid waste after collection and before disposal. Processing consists of but is not limited to volume reduction, separation, storage, exchange, recovery of energy and resources, biological, chemical, physical, or biological modification, and the processes of a salvage facility or metal recycling (Insider, 2013). Waste processing techniques rely on the type of waste material. For example, hazardous waste undergoes three treatment stages and renders them as unhazardous or less hazardous. (1) Physical processes, such as resin or carbon adsorption, flocculation, sedimentation, centrifuging, reverse osmosis, and ultrafiltration; (2) chemical processes, such as neutralization, fixation, ion exchange, precipitation, oxidation-reduction, and biological processes, such as land application, composting, and activated sludge (Kiang & Metry, 1982).

Like any other process, high energy usage leads to high financial costs which later negatively impacts the environment and the economy if the methods are not properly mediated. In the United Kingdom, Rizan and co-workers investigated the carbon footprint of waste streams in a UK hospital, where it was noted that clinical, anatomical, and medicinal waste streams have the highest carbon footprint (Rizan et al., 2021). Plastics play an important role in our society due to their versatility such as shopping, used as bin bags, decoration, and storage. Elsewhere, anaerobic co-digestion of bioplastics as a sustainable mode of waste management with improved energy production was reviewed as another form of waste processing technique (Abraham et al., 2021).

Waste processing is the treatment of solid waste following collection and prior to disposal. Processing includes operations at metal recycling or salvage facilities as well as volume reduction, storage, separation, exchange, energy, and resource recovery, and physical, chemical, or biological modification (Dlamini et al., 2019). A solid waste processing plant, which does not include solid waste transportation, is a facility that alters the chemical, physical, or biological properties of solid waste to facilitate disposal, resource recovery, or transfer (Kurniawan et al., 2022).

- **Stages of Solid Waste Management Process** (Kaza et al., 2018)

  Stage 1: Waste generation, exploitation of natural resources, product manufacturing, product consumption, and material becomes useless and is discarded.

  Stage 2: On-site processing and storage Waste collection, processing (including waste component separation and resource/energy recovery), and storage at or near the point of generation.

  Stage 3: Communal waste storage, garbage collection from sources by external bodies (common in developed nations) and transfer stations or nearby disposal locations, or waste collection by micro-enterprises.

  Stage 4: Transfer station, which is prevalent in low-income nations (for all waste categories), storage in communal bins, containers, or designated locations for solid waste collection by external entities.

  Stage 5: Waste secondary collection and transfer from small collection vehicles to large haulage vehicles.

  Stage 6: Recycling, processing, resource recovery, waste collection from secondary storage location, and transfer to final disposal location.

  Stage 7: Waste disposal and treatment facility is often the last phase of solid waste management.

5.3.3 Resource Recovery

Resource recovery is the practice of using trash as an input material to produce new, valuable products (Nithya et al., 2021). The objective is to decrease or diminish waste production in order to reduce the need for landfill space and increase the value that can be extracted from trash. Due to resource recovery, the use of resources in the manufacturing process is delayed. New materials and products can be made using materials recovered from municipal solid waste, waste from construction and demolition projects, commercial garbage, and industrial waste (Babu et al., 2021). Plastic, paper, aluminium, glass, and metal are just a few of the valuable materials that can be found in waste.

Waste management is just one aspect of resource recovery. Resource recovery is part of a circular economy, which reduces waste production and natural resource extraction and emphasizes the use of materials and products that are more durable, repairable, reusable, and recyclable (Babu et al., 2021). Using life-cycle analysis (LCA), the resource recovery potential of various treatment procedures can be compared. Recovery can be a target in the context of sanitation.

The phrase refers to techniques for recovering the resources present in wastewater and human excreta. In this context, we can talk about urine and faeces, for example (Massoud et al., 2009). Recently, the term "toilet resources" has gained popularity, and the amount of waste that is released into waterways via sewage is rising steadily. These resources include organic matter, energy, water, and nutrients such as nitrogen and phosphorus (Woodard, 2001). Poor sanitation is another name for this core principle. Resource recovery may be helped by separating waste flows. Greywater and blackwater separation, as in urine diversion toilets, and keeping urine and faeces apart, are two examples.

- ❖ **Sources of Resource Recovery**

  Resource recovery can be facilitated by changes in government policy and regulation, circular economy infrastructure, such as enhanced trash cans or bins to encourage source separation and waste collection, reuse, and recycling (Woodard, 2001). Creative circular business models, valuing materials and products not just in terms of their economic costs and benefits but also in terms of their social and environmental costs and benefits (McGinty, 2020). For instance, organic materials can be composted and anaerobically digested to create compost or fertilizer. Similar to this, wastes currently kept in industrial landfills and in closed mines can be processed using bioleaching and specially designed nanoparticles to recover metals like vanadium, cobalt, and lithium for use in low-carbon technologies like wind turbines and electric cars (Fauziah & Agamuthu, 2012).

  A restricting factor in resource recovery is the irreversible loss of raw materials caused by their increase in entropy in our current linear economic paradigm. Entropy is increased in a variety of ways, starting with waste production in manufacturing and continuing through corrosion and wear and tear over the course of usage (Nithya et al., 2021). At the end of the life cycle, an exponential rise in disorder is brought on by the mixing of materials in landfills. The directionality of the entropy law results in a decreasing potential for resource recovery. This supports the need for a circular economy business strategy and infrastructure (Babu et al., 2021).

  Recycling is a technique for recovering resources that entails gathering and using waste products like empty beverage containers (Babu et al., 2021). The components used to create the objects which have become waste can be reused to create new goods. Recycling materials can either be arranged directly from diverse waste streams or collected separately from regular waste using designated bins and collection vehicles (Sulemana et al., 2018). Among the most frequently recycled consumer goods are aluminium beverage cans, polyethylene and PET bottles, copper wire, old steel furniture or equipment, glass bottles and jars, steel food and aerosol cans, paperboard cartons, newspapers, magazines, and light paper, as well as corrugated fibreboards boxes (Mamphitha, 2012).

❖ **Waste treatment technologies.**

1. **Anaerobic digestion (AD)**

   The biological process known as anaerobic digestion involves anaerobic microorganisms breaking down organic matter into smaller compounds (Babu et al., 2021). A number of metabolic phases, including hydrolysis, acidogenesis, acetogenesis, and methanogenesis, make up the AD process. During the hydrolysis stage, expansive organic compounds like proteins, fatty acids, and carbohydrates are broken down into smaller, more soluble molecules like sugars and amino acids (Nascimento et al., 2019). The microorganism consortium of the anaerobic digester produces the enzymes necessary for this transformation. Because the hydrolysis of some multifaceted organic lignocellulosic organisms is so slow, this step in the AD process is regarded as the rate-limiting step (Babu et al., 2021).

   Inorganic compounds like hydrogen sulphide (H_2S), volatile fatty acids (VFA), alcohols, carbon dioxide (CO_2), ammonia (NH_3), and hydrogen are produced from the by-products of hydrolysis during the second stage of acidogenesis (H_2). This transformation is accomplished by an assortment of facultative and obligate anaerobic bacteria. The third phase, known as acetogenesis, results in the production of carbon dioxide, acetic acid, and hydrogen by converting the substances created during the acid genic step (Babu et al., 2021).

   VFAs generated throughout the acidogenesis and acetogenesis steps are essential for nursing and assessing the AD process's stability (Nascimento et al., 2019). The final process in the AD process is methanogenesis, in which waste matter is deteriorated and biogas is generated. This happens when acetotrophic and hydrogenotrophic methanogens are present. Because hydrogen is scarce in AD systems, acetotrophic methanogens produce the vast majority of methane (70%). Acetotrophic methanogens generate methane by breaking down acetate, whereas hydrogenotrophic microbes generate methane by combining hydrogen and carbon dioxide (Babu et al., 2021). The major by-product of the AD process is biogas, which consists primarily of methane (50–75%) and carbon dioxide (Nascimento et al., 2019). The amount of biogas generated is determined by the conformation or arrangement of the waste. Methane can be utilized in conjunction with or instead of traditional cooking fuel (Babu et al., 2021). The creation of biogas is another by-product of the process. It is used as a natural fertilizer because it contains nutrients such as potassium, nitrogen, and phosphorus. The digestion procedure, retention period, and operating temperatures all affect the amount of MSW. Hydrogen is yet another AD by-product made during the process of dark fermentation (Dlamini et al., 2019).

2. **Composting (aerobic digestion)**

   According to Richard et al. (2019), composting is a biochemical process that degrades and stabilizes organic carbon so that it can be used as a

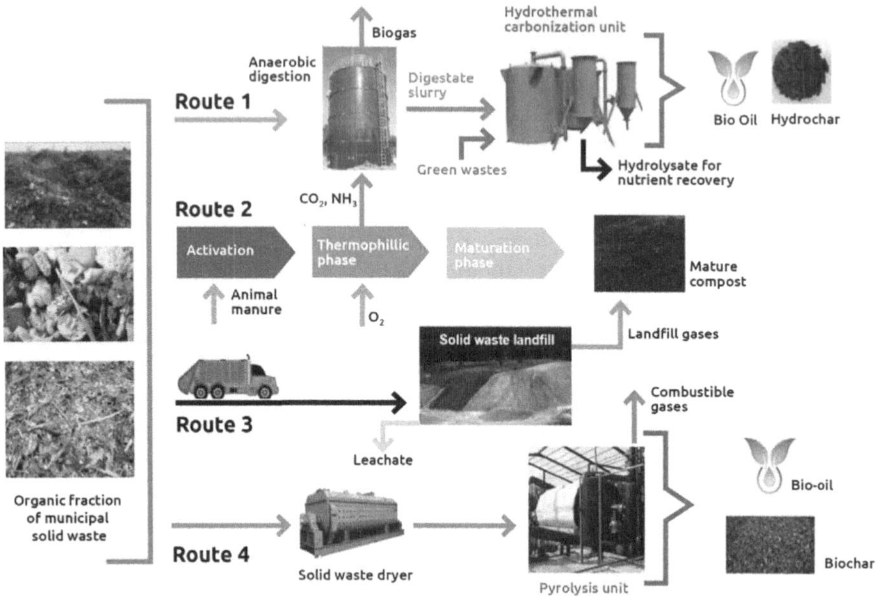

FIGURE 5.8 A flowchart illustrating different treatment methods for converting the organic portion of solid municipal waste into usable resources.

fertilizer or soil conditioner (Route 2, in Figure 5.8). The initial activation, thermophilic, and mesophilic or maturation phases make up the composting process. Mesophilic microorganisms carry out the initial decomposition of compounds while the temperature in the compost pile rises. The thermophilic microorganisms, which can easily thrive in those conditions, take the place of the mesophilic microorganisms as the temperature rises above 40°C. During the thermophilic phase, when microorganisms use oxygen and release carbon dioxide and ammonia, the organic waste decomposes (Babu et al., 2021).

Pathogens can be destroyed, and the viability of seeds present in the organic solid waste can be decreased by temperatures in the range of 55 to 70°C reached during this phase (Babu et al., 2021). Aeration and mixing should be used to keep the temperature below 65°C because temperatures above this point risk killing the microbes that limit the rate of decomposition. As the organic fraction of the compost exhausts, the temperature in the pile drops, ushering in the maturation phase. In this stage, the organic material is stabilized and humidified to create compost with humid properties (Richard et al., 2019).

Compost is the main end result of the composting process. It is a source of organic matter and has qualities like good water retention, buffering abilities, porosity, and structure that make it useful as both a fertilizer and a soil improver (Babu et al., 2021). The composition of the compost is affected by factors like the texture of the raw materials, the size of the particles, the type of inoculum, the moisture content, temperature, pH, aeration time, and heat insulation. Although the composting process occurs naturally, effective composting calls for odour and dust control as well as the acquisition of high-quality agricultural fertilizer that is balanced in N, P, K, and C (Richard et al., 2019).

3. **Hydrothermal carbonization (HTC).**
   In the thermochemical conversion process known as hydrothermal carbonization wet (Dlamini et al., 2019), organic substrates (those with a moisture content of more than 80% by weight) react in hot, pressurized water (Route 1, in Figure 5.8). Typical process conditions range from 10 to 40 bars of pressure and temperatures between 180 and 250 degree Celsius to keep water liquid. The typical residence period lasts anywhere from a few seconds to several hours. Hot pressurized water behaves as an acid/base catalyst precursor and functions as both a solvent and a reactant, exhibiting a higher ion product than at ambient conditions (Babu et al., 2021).
   As a result, the organic matter can decompose and the carbon content rises. The hydrochar, a carbonaceous solid with a carbon content typically greater than 80% by weight, is the main by-product of this process. Its chemical makeup is comparable to that of low-rank coals like lignite or peat (Babu et al., 2021). Applications for hydrochar include use as a solid fuel, a soil enhancer, and an adsorbent.
   The properties of the feedstock and the operating conditions are related to the hydrochar yields and physicochemical properties. The feedstock's polymeric structure, carbon and mineral content, and degradation pathway are all influenced by these factors (Richard et al., 2019). Socioeconomic status, consumption patterns, and collection sources of MSW have an impact on waste composition. Typically, it contains a high proportion of volatile matter and a medium to low amount of fixed carbon. The HTC application enables the conversion of MSW and guarantees that it carbonizes at low temperatures (Babu et al., 2021).
4. **Pyrolysis**
   The process of thermally decomposing organic material without oxygen is known as pyrolysis (Richard et al., 2019). This method is used to turn MSW into biochar. There are two different types of pyrolysis which are according to Babu et al. (2021); slow pyrolysis, which is typically used to increase biochar yield up to 35% and is carried out in the range of 400 – 600 °C with a heating rate around 0.2 C/s, and fast pyrolysis, which is carried out at temperatures higher than 1000 °C with a heating rate of roughly 500 C/s and produces up to 70% yield of bio-oil as a major product (Route 4, in Figure 5.8).

By using the pyrolysis method to convert MSW, three main products can be produced: biochar, a solid that is rich in carbon; bio-oil, a liquid that results from the condensation of volatile organic matter; and non-condensable gases like hydrogen, carbon monoxide, carbon dioxide, and methane factors (Richard et al., 2019). Because of its numerous uses in the areas of wastewater treatment, soil amendment, soil remediation, carbon dioxide capture, and energy storage, biochar has garnered the most attention internationally among these products. The stability of biochar is a crucial component (Babu et al., 2021).

5. **Landfilling**

   Landfilling is an end-of-pipe waste treatment process that entails the decomposition of organic waste and the long-term storage of inert materials as heaps (Dlamini et al., 2019). Organic waste decomposition in a landfill is relatively uncontrolled. For resource recovery purposes, two outputs can be collected which are landfill gas and landfill leachate (Route 3, Route 1, in Figure 5.8). Because of the presence of organic matter in the waste, landfill leachate is a dark liquid. The leachate is high in humic and fulvic acids, as well as volatile organic compounds and persistent organic pollutants (Babu et al., 2021).

   Landfill gas is composed of 50% methane, 50% carbon dioxide, and a trace of non-methane organic compounds. Landfill gas is removed from landfills using a network of wells and pipes. The gas can then be converted into electricity, utilized in a landfill gas power project, and then usefully distributed to the general public such as through grid systems (Babu et al., 2021).

6. **Incineration**

   The most expensive option for managing solid waste is to incinerate municipal solid waste. If incineration is to be used as a waste management method, skilled workers and careful maintenance are required (Ayeleru, 2016). In light of the availability of simpler and less expensive alternatives, incineration might not be the best choice. Setting up an incineration plant requires a large upfront investment and ongoing operating expenses. Due to a lack of landfills in the cities, incineration has replaced landfilling as the primary method of managing MSW in the majority of the world (Iyamu et al., 2020).

   The main goal of using incineration technology for MSW treatment is to reduce waste production to as little as 95% (Iyamu et al., 2020). Because of the threat it poses to both the environment and human health, the incineration of WM has become a contentious disposal method in some nations around the world. The incineration method of waste disposal has fallen out of favour for technical and financial reasons. The implementation of the incineration method does not ensure human health or minimal or no environmental impacts, which are the main concerns of MSWM. Even though the mass and volume can be cut by about 75% and 90% according to (Ayeleru, 2016), respectively, it is still not a solution because solid remnants like fly ashes will remain.

5.3.4 Waste Disposal

Waste can be classified as liquid or solid depending on its source, structure, and effects on the environment and human health (Ayeleru, 2016). When municipal, manufacturing, and other solid waste is treated and disposed of, huge quantities of methane are manufactured (CH_4). In addition to CH_4, solid waste disposal sites (SWDS) also release small traces of nitrous oxide (N_2O), nitrogen oxides (NOx), carbon monoxide (CO), and biogenic carbon dioxide (CO_2), non-methane organic volatile compounds (NMVOCs), and water vapour (Song & Zeng, 2015). About 3 to 4% of annual worldwide anthropogenic releases of greenhouse gases come from solid waste disposal sites (Dlamini et al., 2019).

In the past ten years, waste management has undergone important changes in many developed nations (Achankeng, 2003). Waste minimization and recycling and reuse regulations have been put in place to cut down on waste production, and more and more alternative waste management techniques are being used to lessen the negative environmental effects of solid waste disposal on land. Additionally, waste gas recovery is gaining popularity as a way to lower methane emissions from solid disposal facilities. The decomposition of the organic material that comes from biomass sources is the main source of CO_2 launched from waste (e.g., crops, wood). Since the carbon is biological in nature, it is not counted in national totals of CO_2 emissions (Dlamini et al., 2019).

According to the latest statistics, South Africans generate roughly 122million tons of waste per year. Of this waste, a maximum of only 10% is recycled or recovered for other uses, while at least 90% is landfilled or dumped illegally (Department of Environmental Affairs, 2012; 3SMedia, 2022). Currently, the South African municipalities across different metropolitans are still struggling to mitigate illegal waste disposal practices due to high population. According to reports, for the past 20 years, there were no new licensed landfill sites across different provinces (Department of Environmental Affairs, 2012).

5.4 OPPORTUNITIES TO SHIFT TOWARDS A CIRCULAR ECONOMY

Africa needs to understand the advantages secondary resources offer the continent. Reusing, recycling, and recovering trash instead of disposing of it in landfills and dumps could boost the African economy by $8 billion annually and create a wealth of socioeconomic opportunities (Mitchell & James, 2015). If the objective is accomplished, secondary resources will be brought back into the African economy, increasing and bolstering local manufacturing, generating employment, addressing unemployment, and constructing local and regional economies. Additionally, if completed ethically and sustainably, this will reduce the negative effects of the continent's current subpar solid waste management practices on the environment and human health (Kurniawan et al., 2022). Figure 5. 9 shows the linear and circular economy on solid waste management.

5.4.1 Make Better Use of Finite Resources

Therefore, using natural resources more effectively includes making better use of forests, land, water, air, metals, and minerals (McGinty, 2020). Think about the

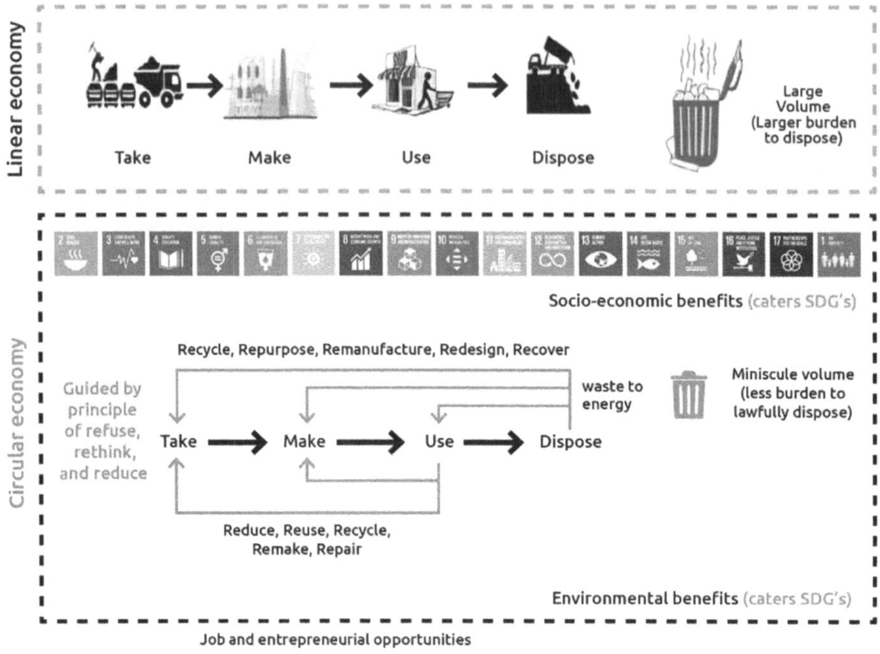

FIGURE 5.9 An image showing linear and circular economy on solid waste management.

textile business. The production of clothing made of synthetic fibres requires enormous quantities of fossil fuels each year. Each year, the production of textiles (including cotton farming) uses about 100 billion cubic metres, or 4% of all freshwaters that are emitted into the atmosphere. At the same time, people throw away still-wearable clothing estimated to be worth $460 billion annually (Kurniawan et al., 2022).

Utilizing recyclable and recycled materials will help create a textile circular economy by lowering the amount of land, water, and fossil fuels required to produce new clothing (McGinty, 2020). It involves changing consumption patterns to reduce new clothing purchases and keep clothing in use for longer periods of time, such as by creating rental and second-hand markets and altering the fast fashion culture. According to research done by Nascimento et al. (2019), acquiring 100 used clothes can replace making 85 brand-new ones. Last but not least, it entails making sure that used clothing is gathered, recycled, or turned into new clothing to further reduce resource use.

5.4.2 Reduce Emissions

Around 45% of global greenhouse gas emissions come from the production, consumption, and manufacture of goods (McGinty & Craven, 2021). Techniques for a circular economy that conserve resources can cut emissions of greenhouse gases by 39% (22.8 billion tonnes) globally and be crucial in preventing the dire effects of climate change (Muscio & Sisto, 2020). For instance, using recyclable materials would completely eradicate the need to create new synthetic fibres and polymers, resulting in significant reductions in the depletion of fossil fuels and the resulting emissions.

Altering consumption rates is also crucial. For instance, the textile industry's greenhouse gas emissions would decrease by 44% if the average amount of times a garment was worn doubled. Developing a circular economy for food by minimizing waste and loss is crucial for lowering emissions. If food loss and waste were a nation, it would come in third behind China and the United States (Fuller et al., 2022).

5.4.3 Protect Human Health and Biodiversity

Air, water, and soil pollution cause more than nine million deaths worldwide each year. According to Fuller et al. (2022), this pollution also poses a hazard to biodiversity. Human health and biodiversity are benefited by working towards a circular economy in a number of ways, which would include better use of land and resources (such as protecting water and land) and averting the climatic catastrophe. How we dispose of items at the end of their useful lives will be one of the most obvious and immediate effects of the shift to a circular economy (Kurniawan et al., 2022).

The weight of the entire human population is roughly equivalent to the 300 million tonnes of plastic waste that the world produces each year (McGinty, 2020). On top of that, only 17.4% of the 54 million tonnes of electronic waste (e-waste) are collected and recycled (Nithya et al., 2021). When waste is improperly managed, it can harm biodiversity and human health by escaping into the environment or by being disposed of in landfills, open fires, or improper recycling.

The amount of waste produced is reduced when products are made to last longer. Creating appropriate collection and processing techniques protects people and the environment from harmful pollutant (Ngoc & Schnitzer, 2009). For instance, implementing currently available strategies like switching out plastic for different materials, designing plastics to be more easily recycled, and stepping up collection and recycling could cut the flow of plastic pollution into the ocean by 80% in 20 years, which would have an important positive impact on human health and biodiversity (Horejs, 2020).

5.4.4 Boost Economies

Muscio and Sisto (2020) conducted a study, and statistics show that the circular economy's ability to reduce waste, foster innovation, and generate jobs opens up a $4.5 trillion economic opportunity. Innovative opportunities are enormously expanded by new business models that emphasize sharing, reusing, repairing, and remanufacturing. For instance, a circular economy for plastics offers considerable financial advantages. Since plastic pollution currently costs and results in economic losses lot annually, less plastic trash in the ocean would benefit sectors like fishing and tourism (Newman et al., 2015). Lowering the use of fossil fuels for plastic production would help mitigate climate change and its costs, while reducing pollution and harmful emissions from open burning of plastic waste would result in lower healthcare costs.

A long-term perspective as well as short-term incentives to encourage change are crucial because many of these economic advantages and opportunities are long-term, indirect, and involve significant investment (Wei et al., 2010). Examples of such policies include those that provide more immediate financial incentives for businesses to create novel new business models and support the effective flow of reused and recycled materials throughout international value chains.

5.4.5 Create More and Better Jobs

According to Mitchell and James (2015), by 2030, shifting to a circular economy might create an additional 1.2 million jobs. It will take a strong commitment to social and environmental justice to fully realize this potential. However, new jobs will be created in industries like recycling, services like repair and renting, or new businesses that emerge to make inventive use of secondary materials. Jobs in more linear sectors may disappear (Alhola et al., 2019). Because they may be in different areas and require different abilities, these new professions cannot be considered direct replacements.

For instance, we must consider the millions of textile workers, most of whom are women, whose livelihoods depend on the survival of the fast fashion industry. Investment in a just transition must be made through social dialogue, social protection, and reskilling programmes to achieve the SDGs. While a net gain in employment is significant, another benefit of circularity is the ability to give formal work and better working conditions for informal labourers. Around 500,000 people all over the globe work as "trash pickers", extracting useful or recyclable items from trash (Mamphitha, 2012). Introducing these unofficial waste pickers into formal jobs in collection or recycling represents a significant chance to provide safer, more secure employment.

5.4.6 Resource Recovery

In order to minimize the amount of waste that is disposed into landfill sites and to reduce the amount of waste generated, waste can be used as an input to generate valuable products (Lacovidou et al., 2017). The necessity for effective MSW management in an economically feasible, environmentally friendly, socially acceptable, and technologically viable way to encounter problems facing developing countries in general. In Western Cape, South Africa a decision-support model for selection of appropriate technologies for optimal resource recovery from different classes of waste was introduced and investigated.

This model ensures environmental and human health protection, supports economic development, and fulfills social and regulatory requirements for reduction of waste that goes to landfills (Adenuga & Popoola, 2020). Biochemical techniques such as composting, vermicomposting, and anaerobic digestion have been investigated for the management of municipal solid waste into value-added products (Varjani et al., 2021). In developed countries, conversion of solid waste to energy is regarded as a feasible choice for the dumping of municipal solid waste and the development of energy.

Waste-generated fuel can have a minor carbon footprint than fossil fuel and also gasification (thermal conversion) of biomass and waste may possibly reduce harmful gases and toxic elements (Rehfeldt et al., 2020; Hameed et al., 2021). Gasification is a powerful technology in waste management which aids in converting municipal solid waste into a useful syngas which contains hydrogen, carbon monoxide, and carbon dioxide. The syngas can be used for the generation of energy by fuel cells, gas turbines, or gas engines (Oh et al., 2018; Veksha et al., 2018). More environmentally friendly and commercially viable technologies need to be looked into in order to mitigate and minimize the levels of solid waste in landfills and also build a viable circular economy.

5.4.7 Maximizing the Impact of the Circular Economy

There are always trade-offs to consider and manage when working towards broad-scale, systemic change. The use of fossil fuels will be reduced by switching to bio-based plastics and natural, recyclable textiles like cotton, for example, as opposed to conventional plastics or synthetic fibres, but the amount of land and water needed to grow these materials may increase (McGinty & Craven, 2021). Shifting to natural materials is an important part of the solution, but only if they are produced in a sustainable manner and if consumer behaviours shift as well.

Recognizing the interconnectivity of the world economy is also fundamental. The extraction of aluminium, copper, lead, and zinc, which are used in a variety of industries, produces many minerals and metals that are required in electronics. Circular manufacturing in the electronics sector would not be sufficient to significantly lessen reliance on such resources (Nascimento et al., 2019). Multiple industries must adapt to this in future in order to bring about systemic change.

The importance of putting social equity and well-being first cannot be overstated. For instance, transitioning to a circular economy might refocus investment and employment away from production and manufacturing which typically take place in low-income countries and towards later stages of the value chain, like repair, resale, sorting, and recycling which are often concentrated in wealthier countries (Corrêa & Xavier, 2013). We must make sure that economic benefits are distributed fairly in order to fully realize the advantages of a circular economy. Figure 5.10 shows the waste management and opportunities that the circular economy presents.

Africa's first objective is to address the need for proper waste management by ensuring that all citizens have access to these services (Massoud et al., 2009). Every integrated waste management system is based on thorough, dependable, and regular garbage removal and city cleaning. According to Li and Achal (2020) all citizens should have access to sufficient waste collection services and eco-friendly waste management by the year 2030.

The second objective is to move waste up the waste management hierarchy, away from disposal and towards waste avoidance, reuse, recycling, and recovery, in order to unlock the socioeconomic potential of trash as a resource (Song et al., 2015). This opportunity is in line with SDG 11, which is all about making cities and human settlements more inclusive, safe, resilient, and maintainable. SDG 13: Take immediate action to mitigate climate change and its consequences. Waste disposal can have

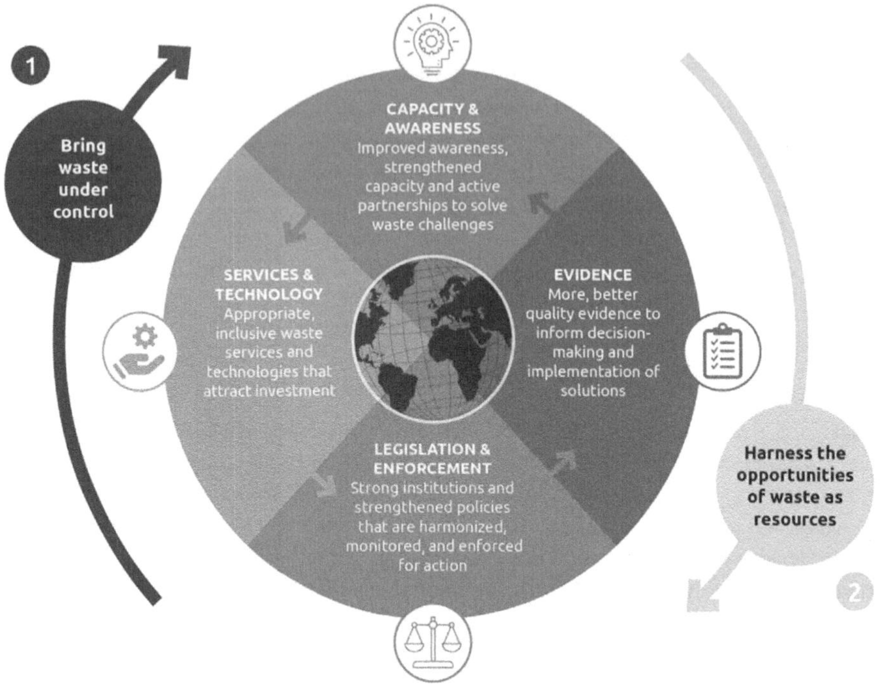

FIGURE 5.10 Circular Economy showing waste management and opportunities.

an impact on climate change. Also included are SDGs 15 and 16. By 2030, African countries will have diverted at least 50% of their garbage to waste reuse, recycling, and recovery, and waste prevention measures are in place (Kaza et al., 2018).

5.5 PROBLEMS

1. What are the typical obstacles that African municipalities encounter when it comes to solid waste management?
2. What are the differences between Africa's current municipal solid waste management system and international best practices?
3. What measures or schemes have African municipalities taken to enhance their management of solid waste?
4. Are there any solid waste management case studies in African municipalities that have been successful enough to be considered models for others?
5. Are the present solid waste management practices in African municipalities linked with any environmental or health apprehensions?
6. In African municipalities, what part does the community have in managing solid waste?
7. What measures are being taken by African municipalities to tackle waste segregation and recycling challenges?

8. Is the employment of inventive technologies for enhancing solid waste management observable in African municipalities?
9. How does solid waste management affect the finances of municipalities in Africa?
10. In what ways could solid waste management practices be enhanced through partnerships and collaboration among African municipalities and other stakeholders?

REFERENCES

3SMedia. (2022). State of the South African waste industry. *Infrastructure News*. https://infrastructurenews.co.za/2022/03/10/state-of-the-south-african-waste-industry/#:~:text=According%20to%20the%20latest%20available,is%20landfilled%20or%20dumped%20illegally

Abraham, A., Park, H., Choi, O., & Sang, B. I. (2021). Anaerobic co-digestion of bioplastics as a sustainable mode of waste management with improved energy production: A review. *Bioresource Technology*, *322*, 124537.

Achankeng, E. (2003). Globalization, urbanization and municipal solid waste management in Africa. *Proceedings of the African Studies Association of Australasia and the Pacific 26th Annual Conference*, 1–22.

Adeleke, O., Akinlabi, S., Jen, T.-C., & Dunmade, I. (2021). Towards sustainability in municipal solid waste management in South Africa: A survey of challenges and prospects. *Transactions of the Royal Society of South Africa*, *76*(1), 53–66.

Adenuga, O. A., & Popoola, O. I. (2020). Subsurface characterization using electrical resistivity and MASW techniques for suitable municipal solid waste disposal site. *SN Applied Sciences*, *2*(9), 1549.

United Nations Environment Programme (2023). *Africa Waste Management Outlook*. International Environmental Technology Centre. Available at: https://www.unep.org/ietc/resources/publication/africa-waste-management-outlook (Accessed: 8 April 2025).

Alhola, K., Ryding, S. O., Salmenperä, H., & Busch, N. J. (2019). Exploiting the potential of public procurement: Opportunities for circular economy. *Journal of Industrial Ecology*, *23*(1), 96–109.

Aliu, I. R., Adeyemi, O. E., & Adebayo, A. (2014). Municipal household solid waste collection strategies in an African megacity: Analysis of public private partnership performance in Lagos. *Waste Management & Research*, *32*(9_suppl), 67–78.

Ayeleru, O. O. (2016). *Characterization, management and utilization of landfill municipal solid waste: A case study of Soweto*. University of Johannesburg (South Africa).

Baawain, M., Al-Mamun, A., Omidvarborna, H., & Al-Amri, W. (2017). Ultimate composition analysis of municipal solid waste in Muscat. *Journal of Cleaner Production*, *148*, 355–362.

Babu, R., Prieto Veramendi, P. M., & Rene, E. R. (2021). Strategies for resource recovery from the organic fraction of municipal solid waste. *Case Studies in Chemical and Environmental Engineering*, *3*. https://doi.org/10.1016/j.cscee.2021.100098

Bello, I. A., bin Ismail, M. N., & Kabbashi, N. A. (2016). Solid Waste Management in Africa: A Review. *International Journal of Waste Resources*.

Bhada-Tata, P., & Hoornweg, D. (2016). Solid waste and climate change. *State of the World*, 239–255. https://doi.org/10.5822/978-1-61091-756-8_20

Brazil (Brasilia). (2010, August 2). *National policy on solid waste—law no 12.305*. Republic Diary.

Carlos-Alberola, M., Gallardo Izquierdo, A., Colomer-Mendoza, F. J., & Barreda-Albert, E. (2021). Design of a municipal solid waste collection system in situations with a lack of resources: Nikki (Benin), a case in Africa. *Sustainability*, *13*(4), 1785.

CEMPRE. (2019). *Corporate commitment to recycling.*

Christensen, T. (Ed.). (2011). *Solid waste technology and management.* John Wiley & Sons.

Corrêa, H. L., & Xavier, L. H. (2013). Concepts, design and implementation of reverse logistics systems for sustainable supply chains in Brazil. *Journal of Operations and Supply Chain Management*, *6*(1), 1–25.

Damanhuri, E., Handoko, W., & Padmi, T. (2014). Municipal solid waste management in Indonesia. *Municipal Solid Waste Management in Asia and the Pacific Islands: Challenges and Strategic Solutions*, 139–155.

DEA. (2018). *South Africa state of waste report.* http://sawic.environment.gov.za/documents/8635.pdf

Department of Environmental Affairs. (2012). *National waste information baseline report.* Republic of South Africa, Department of Environmental Affairs. http://sawic.environment.gov.za/documents/1880.pdf

Dlamini, S., Simatele, M. D., & Serge Kubanza, N. (2019). Municipal solid waste management in South Africa: From waste to energy recovery through waste-to-energy technologies in Johannesburg. *Local Environment*, *24*(3), 249–257.

Fauziah, S. H., & Agamuthu, P. (2012). Trends in sustainable landfilling in Malaysia, a developing country. *Waste Management & Research*, *30*(7), 656–663.

Fuller, R., Landrigan, P. J., Balakrishnan, K., Bathan, G., Bose-O'Reilly, S., Brauer, M., Caravanos, J., Chiles, T., Cohen, A., Corra, L., & Cropper, M. (2022). Pollution and health: A progress update. *The Lancet Planetary Health*, *6*(6), e535–e547.

Godfrey, L., Ahmed, M. T., Gebremedhin, K. G., Katima, J. H., Oelofse, S., Osibanjo, O., Richter, U. H., & Yonli, A. H. (2019). Solid waste management in Africa: Governance failure or development opportunity. *Regional Development in Africa*, *235*.

Gumbo, T., & Simelane, T. (2015). Innovations in municipal solid waste management. *Future Directions of Municipal Solid Waste Management in Africa*, 202–217. https://doi.org/10.2307/j.ctvh8r2sj.18

Hameed, Z., Aslam, M., Khan, Z., Maqsood, K., Atabani, A. E., Ghauri, M., Khurram, M. S., Rehan, M., & Nizami, A. S. (2021). Gasification of municipal solid waste blends with biomass for energy production and resources recovery: Current status, hybrid technologies and innovative prospects. *Renewable and Sustainable Energy Reviews*, *136*, 110375.

Harts, C., & Smith, C. (2008). *Namibia's greenhouse gas inventory for year 2000.* https://unfccc.int/sites/default/files/national_ghg_inventory_report_republic_of_namibia_27_sep_15_final_version_%28002%29.pdf

Hoornweg, D., & Bhada-Tata, P. (2012). *What a waste: A global review of solid waste management.* Urban Development & Local Government Unit, World Bank.

Horejs, C. (2020). Solutions to plastic pollution. *Nature Reviews Materials*, *5*(9), 641.

Imam, A., Mohammed, B., Wilson, D. C., & Cheeseman, C. R. (2008). Solid waste management in Abuja, Nigeria. *Waste Management*, *28*(2), 468–472.

Insider, L. (2013). *Waste processing definition.* https://www.lawinsider.com/dictionary/waste-processing

Iyamu, H. O., Anda, M., & Ho, G. (2020). A review of municipal solid waste management in the BRIC and high-income countries: A thematic framework for low-income countries. *Habitat International*, *95*, 102097.

Kaza, S., Yao, L., Bhada-Tata, P., & Van Woerden, F. (2018). *What a waste 2.0: A global snapshot of solid waste management to 2050.* World Bank Publications.

Kiang, Y. H., & Metry, A. A. (1982). *Hazardous waste processing technology.* Ann Arbor Science Publishers, Inc.

Komen, K., Mrembu, N., & van Niekerk, M. (2016, October). The role of socio-economic factors, seasonality and geographic differences on household waste generation and composition in the city of Tshwane. In *Proceedings of the 23rd WasteCon conference* (pp. 17–21). Johannesburg South Africa.

Korfmacher, K. S. (1997). Solid waste collection systems in developing urban areas of South Africa: An overview and case study. *Waste Management & Research*, *15*(5), 477–494.

Kurniawan, T. A., Liang, X., O'Callaghan, E., Goh, H., Othman, M. H. D., Avtar, R., & Kusworo, T. D. (2022). Transformation of solid waste management in China: Moving towards sustainability through digitalization-based circular economy. *Sustainability*, *14*(4), 2374.

Lacovidou, E., Millward-Hopkins, J., Busch, J., Purnell, P., Velis, C. A., Hahladakis, J. N., Zwirner, O., & Brown, A. (2017). A pathway to circular economy: Developing a conceptual framework for complex value assessment of resources recovered from waste. *Journal of Cleaner Production*, *168*, 1279–1288.

Lema, G., Mesfun, M. G., Eshete, A., & Abdeta, G. (2019). Assessment of status of solid waste management in Asella town, Ethiopia. *BMC Public Health*, *19*(1), 1261.

Li, W., & Achal, V. (2020). Environmental and health impacts due to e-waste disposal in China – A review. *Science of the Total Environment*, *737*, 139745.

Loukil, F., & Rouached, L. (2020). Waste collection criticality index in African cities. *Waste Management*, *103*, 187–197.

Mália, M., De Brito, J., Pinheiro, M. D., & Bravo, M. (2013). Construction and demolition waste indicators. *Waste Management & Research*, *31*(3), 241–255.

Mamphitha, D. (2012). *The role played by subsistence waste pickers in recycling* [Doctoral dissertation, University of Pretoria].

Massoud, M. A., Tarhini, A., & Nasr, J. A. (2009). Decentralized approaches to wastewater treatment and management: Applicability in developing countries. *Journal of Environmental Management*, *90*(1), 652–659.

McGinty, D. (2020). How to build a circular economy. *World Resources Institute*. Retrieved June 29, 2024, from https://www. wri. org/insights/how-build-circular-economy

McGinty, D. B., & Craven, P. (2021). *Opportunities of a circular economy*. World Resources Institute.

Mitchell, P., & James, K. (2015). *Economic growth potential of more circular economies*. Waste and Resources Action Programme (WRAP).

Mochache, M., Yegon, R., & Wakindiki, I. I. C. (2020). Market town household solid waste management: A case study of Embu, Kenya. *Journal of Applied Sciences and Environmental Management*, *24*(1).

Muscio, A., & Sisto, R. (2020). Are agri-food systems really switching to a circular economy model? Implications for European research and innovation policy. *Sustainability*, *12*(14), 5554.

Nascimento, D. L. M., Alencastro, V., Quelhas, O. L. G., Caiado, R. G. G., Garza-Reyes, J. A., Rocha-Lona, L., & Tortorella, G. (2019). Exploring Industry 4.0 technologies to enable circular economy practices in a manufacturing context: A business model proposal. *Journal of Manufacturing Technology Management*, *30*(3), 607–627.

Newman, S., Watkins, E., Farmer, A., Brink, P. T., & Schweitzer, J. P. (2015). The economics of marine litter. *Marine Anthropogenic Litter*, 367–394.

Ngoc, U. N., & Schnitzer, H. (2009). Sustainable solutions for solid waste management in Southeast Asian countries. *Waste Management*, *29*(6), 1982–1995.

Nithya, R., Sivasankari, C., & Thirunavukkarasu, A. (2021). Electronic waste generation, regulation, and metal recovery: A review. *Environmental Chemistry Letters*, *19*, 1347–1368.

Oh, J., Wang, J., & Wiens, J. (2018). Learning to exploit invariances in clinical time-series data using sequence transformer networks. *Proceedings of the 3rd Machine Learning for Healthcare Conference, in Proceedings of Machine Learning Research*, *85*, 332–347. https://proceedings.mlr.press/v85/oh18a.html

Okot-Okumu, J. (2012). Solid waste management in African cities—East Africa. *Waste Management—An Integrated Vision* [Preprint]. https://doi.org/10.5772/50241

Oteng-Ababio, M., Arguello, J. E. M., & Gabbay, O. (2013). Solid waste management in African cities: Sorting the facts from the fads in Accra, Ghana. *Habitat International, 39*, 96–104.

Rasmeni, Z. Z., & Madyira, D. M. (2019). A review of the current municipal solid waste management practices in Johannesburg City townships. *Procedia Manufacturing, 35*, 1025–1031.

Rehfeldt, M., Worrell, E., Eichhammer, W., & Fleiter, T. (2020). A review of the emission reduction potential of fuel switch towards biomass and electricity in European basic materials industry until 2030. *Renewable and Sustainable Energy Reviews, 120.*

Richard, E. N., Hilonga, A., Machunda, R. L., & Njau, K. N. (2019). A review on strategies to optimize metabolic stages of anaerobic digestion of municipal solid wastes towards enhanced resources recovery. *Sustainable Environment Research, 29*, 1–13.

Rizan, C., Bhutta, M. F., Reed, M., & Lillywhite, R. (2021). The carbon footprint of waste streams in a UK hospital. *Journal of Cleaner Production, 286*, 125446.

Song, Q., Li, J., & Zeng, X. (2015). Minimizing the increasing solid waste through zero waste strategy. *Journal of Cleaner Production, 104*, 199–210.

Sulemana, A., Donkor, E. A., Forkuo, E. K., & Oduro-Kwarteng, S. (2018). Optimal routing of solid waste collection trucks: A review of methods. *Journal of Engineering, 2018.*

UN-Habitat. (2010). *Solid waste management in the world's cities: Water and sanitation in the world's cities 2010.* Routledge.

Varjani, S., Shah, A. V., Vyas, S., & Srivastava, V. K. (2021). Processes and prospects on valorizing solid waste for the production of valuable products employing bio-routes: A systematic review. *Chemosphere, 282*, 130954.

Veksha, A., Giannis, A., Oh, W. D., Chang, V. W. C., Lisak, G., & Lim, T. T. (2018). Catalytic activities and resistance to HCl poisoning of Ni-based catalysts during steam reforming of naphthalene. *Applied Catalysis A: General, 557*, 25–38.

Wei, M., Patadia, S., & Kammen, D. M. (2010). Putting renewables and energy efficiency to work: How many jobs can the clean energy industry generate in the US? *Energy Policy, 38*(2), 919–931.

Woodard, F. (2001). *Industrial waste treatment handbook.* Elsevier.

Zhou, M. H., Shen, S. L., Xu, Y. S., & Zhou, A. N. (2019). New policy and implementation of municipal solid waste classification in Shanghai, China. *International Journal of Environmental Research and Public Health, 16*(17), 3099.

6 Mining Waste and Industrial By-products in the African Context

Thandiwe Sithole and Tebogo Mashifana

6.1 INTRODUCTION AND SOURCES

Waste is regarded as something that has no use or value to anyone (Suresh & Vijayakumar, 2012). However, in reality, wastes possess the capability to be used as raw material for various applications (Suresh & Thesis, 2011). Improper handling of wastes could result in several environmental hazards, such as air pollution, soil erosion, and methane emission (Mahla et al., 2013). In developing African countries, appropriate measures for handling waste materials are of great importance, from the environmental prospect. In 2010, Kgosiesele et al. reported that solid waste generation has intensified in many developing countries due to the expansion of technology and industrial processes (Mahla et al., 2013).

African countries have abundant mineral sources such as gold, silver, copper, uranium, and cobalt (Waste Management, n.d.). Moreover, 42 out of 54 countries in Africa are resource dependent while the other 10 are dependent on agriculture (Draft, 2017). The mining sector is crucial to drive economic growth and create employment opportunities. Studies have shown that Africa produces 80% and 50% of the total world supply of platinum and manganese, respectively (Signé & Johnson, 2021; International Institute for Sustainable Development, n.d.). Mining involves the extraction of target commodities by movement and processing of materials at a large scale (International Institute for Sustainable Development, n.d.). During mining activities, the material found in excess is called mine waste (International Institute for Sustainable Development, n.d.; Kamunda et al., 2016). Many mining operations yield waste materials at a large scale containing reactive minerals (International Institute for Sustainable Development, n.d.). As a result, the reactive minerals are released from the mine rock and get discharged into air and water streams. Figure 6.1 shows the various types of waste generated globally in kilograms per person per day. The most common mining wastes are such as slag, tailings, overburden, waste rock piles, and spent heap leach (Doka, 2008; International Institute for Sustainable Development, n.d.; Nwaila et al., 2021). Industrial wastes include waste materials from housekeeping wastes, building wastes, hazardous wastes, and chemical plants wastes as well as commercial garbage from retailer and hospitality industries (DEFF, 2020). Industrial wastes are most commonly in the form of ash, metals, wood, glass, and food wastes (Adler et al., 2007; Draft, 2017). Figure 6.2 shows the average steel and iron process wasted generation to crude carbon steel production from 2008 to 2013.

DOI: 10.1201/9781003377856-6

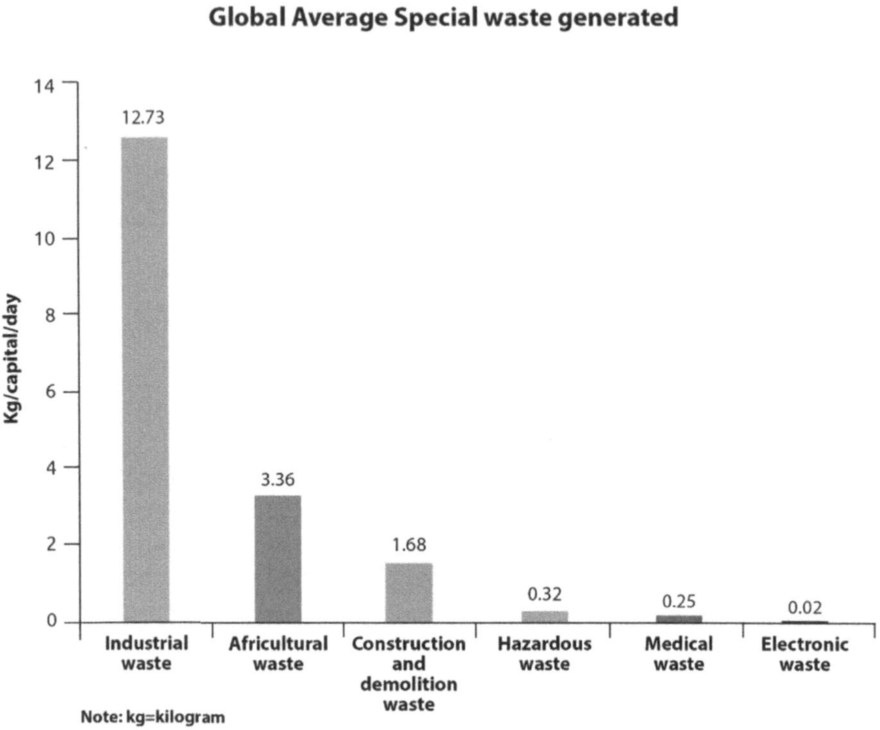

FIGURE 6.1 Various types of waste generated globally in kilograms per person per day.

Source: World Bank, 2023.

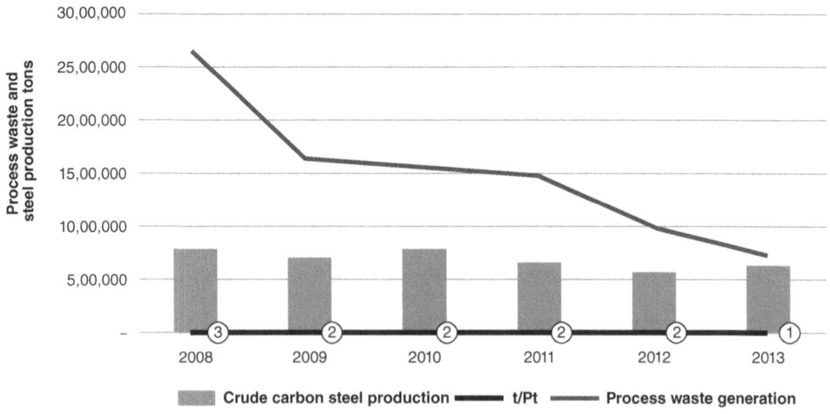

FIGURE 6.2 Average steel and iron process waste generation compared to crude carbon steel production from 2008 to 2013 in Southern Africa.

Source: Schoeman et al., 2021.

6.2 MANAGEMENT PRACTICES CASE STUDIES

6.2.1 Introduction

Wastes have a negative impact on the environment and social lives (Adler et al., 2007; Matsumoto et al., 2016). The government must play a significant role in ensuring that efficient and safe waste management strategies are implemented. Improperly handled wastes result in several environmental hazards, such as air pollution, soil erosion, and methane emission (Signé & Johnson, 2021). As a result, mining companies must exploit many resources and more time to manage mining wastes efficiently. This can be achieved through overburden, waste rock piles, spent heap leach, and facilities tailings management facilities (Nwaila et al., 2021). In mining, the main objective of waste management is to decrease waste-producing volumes.

Currently, several African countries focus on waste management systems such as waste legislation, waste collection, treatment, and disposal methods (Hansen, 2015; Suresh & Thesis, 2011). Waste management strategies assist in ensuring that mining waste and its management facilities are physically and chemically stable for the long term. Innovative ideas and objectives such as repurposing, reprocessing, and reduction of liability must be considered for managing mining wastes (International Institute for Sustainable Development, n.d.). These objectives are in support of the United Nations SDGs 6, 8, 13, 14, and 15 (International Institute for Sustainable Development, n.d.; Signé & Johnson, 2021). This is because stable and small mine waste facilities will ensure the protection of water resources, aquatic life, and life on land (International Institute for Sustainable Development, n.d.). Moreover, waste management involves processes such as waste processing, resource recovery, and waste disposal as represented in Figure 6.3. The most common methods for

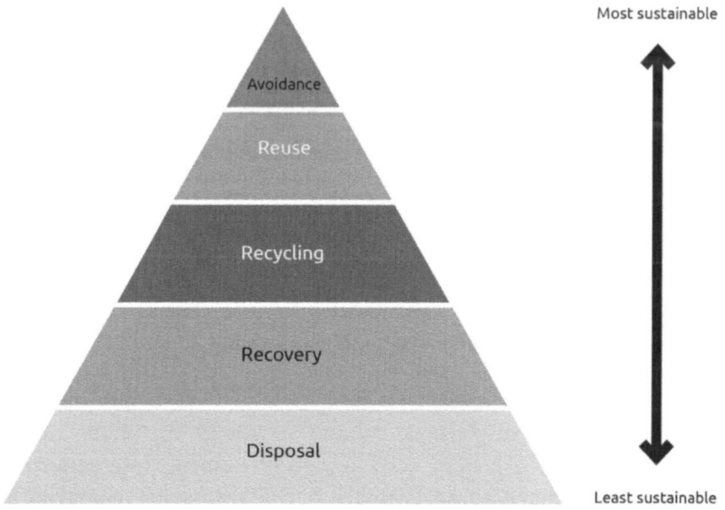

FIGURE 6.3 The waste hierarchy, with waste disposal in the landfill as the least preferred route and avoidance as the most sustainable option in managing waste.

Source: Suresh & Thesis, 2011.

waste management are such as recycling, incineration, and chemical and biological treatment (Silva et al., 2017). Industrial solid waste disposal methods in African countries involve segregation and recycling, use of landfills as well as composting (Onipede & Bolaji, 2004). Segregation is the separation of waste by type (Peša, 2022). In mining industries, waste materials are normally disposed of in landfills (Onipede & Bolaji, 2004).

Waste minimization or avoidance is mostly preferred route in the waste hierarchy because it avoids impacts of the product during its life cycle. Reuse is the utilization of waste products without making any changes or processing them. Reuse is more dominant than recycling in the waste hierarchy since it does not need any new or additional resources for further usage of the product. Reuse decreases the overall cost of developing products, hence favourable to the environment by reducing the need for new raw material. Examples of reused materials are such as plastics, clothes, furniture, and home appliances which have not reached their end of life (Suresh & Thesis, 2011).

Recycling involves the conversion and reprocessing of discarded waste products from a waste stream into raw materials to form the same product or a new product. Recycling includes waste collection, segregation, reprocessing, and manufacturing. The most commonly used waste or by-products for recycling include glass bottles, paper, plastics, and metal cans (DEFF, 2020).

Recovery describes material recovery as well as resource recovery. Material recovery involves the recovery of materials which is further used for recycling. Resource recovery usually involves the conversion of discarded materials into a useful resource either through biological or thermal treatment. Composting, anaerobic digestion, and closed composting are the most common biological processes used for resource recovery. Thermal processes such as pyrolysis and incineration are currently used for resource recovery (Qadir et al., 2010).

Disposal is the least preferred and least sustainable option for waste management. Its main purpose is to remove unwanted material from human settlements. The frequent disposal methods practiced in African countries are landfilling, open dumping, and incineration (Onipede & Bolaji, 2004; Suresh & Thesis, 2011).

In literature, vital information has been provided regarding waste management practices in African countries and their measures towards sustainability. Hence this work focuses on case studies for waste management in the African context. Case studies for Ghana, South Africa, Namibia, Nigeria, and Botswana are outlined in the subsequent sections.

6.2.2 Ghana Management Practices Case Studies

6.2.2.1 Introduction

Ghana has a long history of artisanal mining. In the 1980s, Ghana started large-scale mining activities which came with foreign investment (International Institute for Sustainable Development, n.d.). Ghana is one of the leading countries in Africa for gold production. Furthermore, it produces bauxite, diamonds, and manganese (International Institute for Sustainable Development, n.d.). The International Council on Mining and Metals, 2020 reported that 8.5% contribution of Ghana's gross domestic product (GDP) in 2018 was from the production of metallic minerals. In 2021, Ghana had 16 major operating mines with almost one million people involved in artisanal mining. Ghana has 21 tailings storage facilities for waste management, which present environmental risks. The government of Ghana demonstrates leadership in waste management across Africa. As a result, mining companies in Ghana usually spend more time and resources to manage waste effectually, through waste rock piles, overburden, tailings management facilities, and spent heap leach facilities. Mine waste in Ghana is managed using the following main pieces of legislation represented below in Table 6.1 (Signé & Johnson, 2021).

Reliable national data on waste generation and composition to effectively plan waste management in Ghana is absent.

6.2.3 Case Studies of Management Practices in South Africa

In South Africa, gold is one of the key commodities extracted. Gold contributes positively to the South African economy (Kamunda et al., 2016). Contrarily, it generates large volumes of waste, such as overburden, waste rock, and gold tailings. Furthermore, South Africa is one of the leading countries with offshore mining of seabed (IFC, n.d.). As presented in Table 6.2, industrial waste contributes to 100% of waste disposed in South Africa with 0% percent of waste recycled.

As shown in Figure 6.4, the majority of waste produced in South Africa is from Fly ash accounting for 64% of waste generated in the year 2017. Fly ash has the disadvantage of being radioactive (Petrik et al., 2003).

6.2.3.1 Waste Processing

In South Africa, mine tailings can be processed and used to make stone paper, mining backfill, cement, and bricks (Kamunda et al., 2016). Tailings can also be used for tourism and recreational purposes (Matsumoto et al., 2016). Furthermore, they can be used to generate clean energy. In 2015, Malatse and Ndlovu investigated

TABLE 6.1
Ghana Management Practices Legislation

Year	Legislation
1992:	Ghana's Constitution is for human rights and environmental protection policy and mining waste management.
1993	The Local Government Act (Act 462) is for regional land-use planning and environmental Management.
2006	The Whistleblower Act (Act 720) protects individuals from disclosing corrupt, unlawful, and illegal conduct
2006	The Minerals and Mining Act necessitates the acquirement of permits before mining
2013	The Commission on Human Rights and Administrative Justice Act Is for human rights protection in Ghana.
2015	The Minerals and Mining (Amendment) Act involves paying penalties for illegal small-scale mining.

TABLE 6.2
General Waste Management Option in 2017 (Western Cape Government, 2018).

Waste Type	Local generation	Imports	Exports	Total tonnage managed	Storage/ stockpiles	Recycling/ recovery	Treatment	Disposal
Municipal waste	4821430			4821430	0.0%	0.0%	0.0%	100.0%
Commercial and industrial waste	360884			481179	0.0%	0.0%	0.0%	100.0%
Fly ash and dust	4346080			4346080	0.0%	3.1%	0.0%	96.9%
Bottom ash	6489080			6489080	0.0%	3.1%	0.0%	96.9%
Slag	4859025			4859025	0.0%	0.0%	0.0%	100.0%
Organic waste	19247851	4048	298	19251600	0.0%	49.2%	0.2%	50.8%
Construction and demolition waste	4482992			4482992	0.0%	52.0%	0.0%	48.0%
Paper	2211225	57855	129374	2139706	0.0%	58.0%	0.0%	42.0%
Plastic	1113362	6748	20856	1099254	0.0%	43.7%	0.0%	56.3%
Glass	2752636	38378	11	2791003	0.0%	71.2%	0.0%	28.8%
Metals	4035929	24168	527037	3533059	0.0%	80.0%	0.0%	20.0%
Tyres	174640		12473	162167	76.4%	23.6%	0.0%	0.0%
Other	729615			729615	0.0%	9.0%	0.1%	91.0%
TOTAL	55624749	124449	690 050	52395187	0.2%	34.5%	0.1%	65.2%

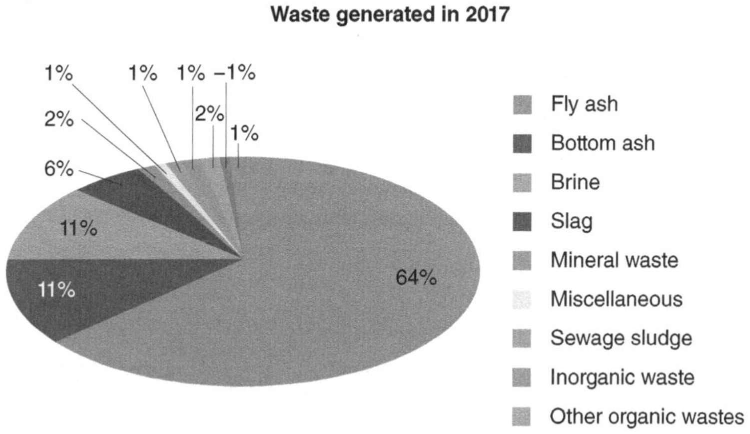

FIGURE 6.4 The percentage of hazardous wastes produced in 2017 in South Africa.

whether it is viable to use gold tailings in South Africa to form bricks (Nwaila et al., 2021). Currently, the utilization of gold tailings in brickmaking is under review and implemented commercially. In the United States, gold tails have been widely used to make ceramics. However, in South Africa, there are no reports of using gold tailings for ceramics making. The main drawback of tailings is that they are radioactive. In cement making, fly ash is common as an additive and tailings have been used as aggregate in construction; its usage for making plaster sand and paving materials (Bian et al., 2012; Nwaila et al., 2021).

Gold tailings can also be used as backfill (Bian et al., 2012; Festin et al., 2019). Currently, there are no academic research findings on the utilization of tailings for backfill. However, many South African mines presently use the backfill for structural stability to avoid rock falls (Bian et al., 2012). Gold Field Mine is one of the South African mines using backfill for structural support (Signé & Johnson, 2021). Harmony's target mine also uses a mixture that contains cyclone tailings additives and cement to make backfill (Bian et al., 2012).

Reprocessing of tails is useful for making stone paper, also known as mineral paper (Zvarivadza, 2018). It is formed by using 80% of tailings as substrate and 20% of polymer acting as binding agent. In 2015, South African media reported that the City of Johannesburg has been exploring tailings together with rubble from illegal dumping areas to make stone paper.

6.2.3.2 Resource Recovery

Gericke et al. (2022) reported that South Africa has developed technology to recover gold and other ores from gold tailings and low-grade ores. Table 6.3 shows the common methods of resource recovery.

6.2.3.3 Waste Disposal

Tailings from South African mines and other industrial wastes are placed in dumps, which cover a larger area of land, approximately 400 square kilometres (Adler et al.,

TABLE 6.3
Showing common methods of resource recovery.

Type of Waste	Recovery method
Gypsum waste	• Solvent extraction • Acid leaching
Metallurgical waste	• Hydrometallurgical • Electrometallurgical
Steel slag	• Dry granulation • Air blast granulation • Granulation with liquid slag impact

2007; Zvarivadza, 2018). Gold tailings are normally found in proximity to residential areas. This poses health risks to humans since mine dumps consist of very reactive and carcinogenic metals such as uranium. Mine dumps also generate acid rock drainage, formed as a result of sulphide exposure to air and water. An alternative approach to land disposal of mine wastes is through repurposing, which is the reallocating of waste feedstock for other applications (Nwaila et al., 2021).

6.2.4 Namibia Management Practices Case Studies

6.2.4.1 Introduction

The mining sector is of great importance to the economy of Namibia (Draft, 2017). It is the main contributor to the country's earnings through foreign exchange. The benefits of mines are that they create employment, thus supplying goods and services. As a result, mining training programs in Namibia contribute effectively towards skill development. In Namibia, there are different types of mines, such as world-class, open-pit, and underground mines (IFC, n.d.). Namibia and South Africa are the leading countries with offshore mining of seabed (Bian et al., 2012). The main commodities mined in Namibia are such as diamonds, copper, uranium, and gold. According to Act, Act 33, 1992 of mineral, mining companies in Namibia are obliged to Namibia's Environmental Plans and Environmental Assessment Policy to prepare for Environmental Assessments (EAs) and Environmental Management Plans (EMPs). This Act is administered by the Ministry of Mines and Energy (MME) to control activities associated with mining. However, artisanal mines in Namibia still lack the background and finance to reduce their mining activities impacts (International Institute for Sustainable Development, n.d.).

6.2.4.2 Industrial Waste

Ongoing implementation of environmental requirements in licenses and specified in the environmental management plans of EIAs. Inventory organizations must develop different types of hazardous waste management so that waste generators are aware of

FIGURE 6.5 Open dumps and vehicles that transport waste to the dumping area.

Source: Ortega et al., 2022.

the options. Furthermore, in the long term, the development of centralized facilities for hazardous waste incinerators is still required in Namibia.

6.2.4.3 Waste Disposal

Many disposal sites in Namibia have been operating for many years, and a large proportion did not have an Environmental Impact Assessment (EIA) carried out before they started operation as shown in Figure 6.5. Even for those that might have had an EIA, reports indicated that most sites do not appear to be properly implementing Environmental Management Plans. Figure 6.5 shows the open dumps and vehicles that transport waste to the dumping area.

6.2.5 Botswana Management Practices Case Studies

Botswana is one of the fastest-growing African countries. It was one of the poorest countries when it got its independence, but it has grown to the rank of middle-income countries within a short span of time. The Republic of Botswana is facing a lot of environmental problems due to its massive development. One of the challenges in Botswana is the solid waste generated in the country. The threats of uncontrolled landfilling are not realized by the people of Botswana. Careless dumping of waste and widespread littering has become a serious problem in Botswana. The growing population in Botswana has resulted in consumption of increased use of products and services which in turn generate a lot of waste (Assfour et al., 2011). The amount of waste generated is closely related to the socioeconomic development of any country (Hogland et al., 2006). In Botswana, the important driver for the generation of solid waste is also found to be socioeconomic development (Ortega et al., 2022). The consumption pattern in Botswana is found to be linear where all the products, which are produced from the natural resources end up as waste and are finally dumped into the landfill without any recovery or recycling activities (Botswana Selected Environmental Indicators Digest 2020, n.d.). To minimize the flow of solid waste ending up in the landfill, the concept of closing the material flows should be taken into account. Closing the material flows involves directing the material flow of virgin materials to used materials (Suresh & Thesis, 2011). Closing the material flows helps to minimize the solid waste ending up in the landfill through reduction, reuse, and recycling activities (Assfour et al., 2011).

As shown in Figure 6.6, general wastes constitute more waste generation (52%). Commercial and industrial wastes contributed 22% of total waste generated in Botswana in 2017.

6.2.5.1 Waste Processing

In Gaborone, metal cans are collected by a company called Collect-A-Can. The recovered metal cans from society are then processed by crushing and pressurization as shown in Figures 6.7A and 6.7B, and then sent to South Africa to further process and recycle the material. Somarelang Tikologo is also one of the NGOs playing a vital role in waste management practices in Botswana and Gaborone. This NGO was founded in 1992 (Ortega et al., 2022). It focuses on four main activities, which involve waste management, environmental planning and public relations, natural resource conservation, and fundraising. The waste management initiatives involve workshop activities, which aim to educate and create awareness among the public regarding the recovery and recycling of waste resources based on the content of the National Waste Management Act. Waste collected by this organization as depicted in Figures 6.7C and 6.7D.

6.2.5.2 Waste Disposal

Disposal in the form of landfilling is still the best practice followed in Gaborone (Botswana Selected Environmental Indicators Digest 2020, n.d.; Suresh & Thesis, 2011). The solid wastes generated throughout Gaborone are buried in the Gamodubu landfill which is owned and operated by the Kweneng district council. Previously, the wastes were disposed in the Gaborone landfill which was later closed during the year 2009 due to the landfill reaching its end of life. From 2009, the wastes were sent to the Gamodubu landfill of up to 76 Hectares with its operation period estimated to be

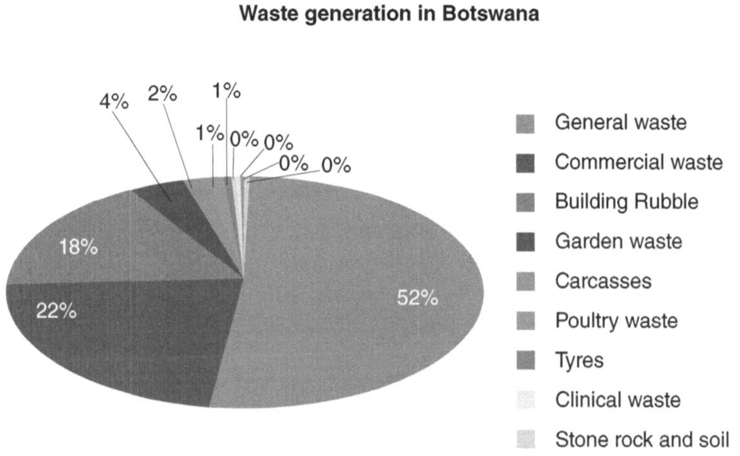

FIGURE 6.6 The amount of waste generated in Botswana in 2017.

Source: Botswana Selected Environmental Indicators Digest 2020, n.d.

FIGURE 6.7 The crushed and pressurized metal can for recycling.

Source: Ortega et al., 2022.

FIGURE 6.8 The weigh bridge in the landfill.

20 years (Suresh & Thesis, 2011). The wastes can be dumped to a height of 5 metres above the ground level which is then covered with soil and surfaced. The Gamodubu landfill was constructed in compliance with health and environmental guidelines. Figure 6.8 shows a weigh bridge in the landfill.

6.2.6 Nigeria Management Practices Case Studies

6.2.6.1 Introduction

In the past years, Nigeria has witnessed urban and economic growth as well as technological development which causes environmental pollution. Industrial activities in Nigeria have resulted in industrial growth, as a result leading to an increase in industrial waste production. The wastes from industrial processes are such as solids, liquids, gases, and air-borne particulate matter, thus get released into the environment. Wastes in Nigeria are government by legislation such as the passage of the Solid Waste Act of 1965.

In 2019, Ogwueleka reported that waste generation in Nigerian urban and rural areas was 0.66 kg/cap/day and 0.44 kg/cap/day, respectively. The increase in waste production in Nigeria is mainly due to expanding population and urbanization. Figure 6.9 shows the composition of waste generated in Lagos.

6.2.6.2 Waste Processing and Resource Recovery

Recycling of wastes through resource recovery attains much potential for the economic growth of a country. Reports have shown that there are no Material Recovery Facilities (MRF) in Nigeria. Moreover, no recycling or resource recovery programs have been established. Organic wastes contribute 60% of total waste generated in the country, with only 8% of it being recovered for reuse.

6.2.6.3 Waste Disposal

Most developing countries practice open dumps waste disposal and Nigeria is also on the list. In Nigeria, there are plenty of uncontrolled landfills located alongside major roads thus causing more traffic. Frequently, the dumped waste is burnt openly

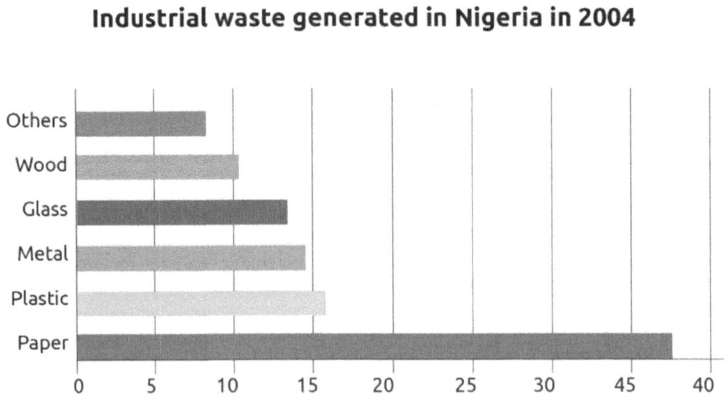

FIGURE 6.9 Waste generated percentage composition from a tipping site located at Isolo, Lagos.

Source: Onipede & Bolaji, 2004.

thus resulting in air pollution. There are no landfill standards and regulations providing basic compliance and waste monitoring in Nigerian landfills. In breweries, spent grains are stored in storage silos and disposed of at dumping sites. The liquid wastes from all sources pass through a network of drains within the factory into a main channel, where they are discharged untreated into the streams or water bodies. The textile industries at Lagos discharge their liquid wastes into the environment untreated. These industries have employed private contractors for the disposal of their solid wastes. In the paper industry, solid wastes from the processes are dumped openly outside the factories. All different types of solid waste are piled up behind the factory to form an unsanitary landfill.

6.2.6.4 Sustainable and Circular Process Engineering Opportunities

The circular economy is an economy that is restorative and regenerative by design and aims to keep products, components, and materials at their highest utility and value always, distinguishing between technical and biological cycles. Various African countries have initiatives for recycling glass, metals, and paper. There is great progress in formalizing the recycling of industrial and mining wastes in many African countries. Numerous African countries have initiated the usage of organic waste for composting and biogas generation, which is expected to increase soon. Research and development on biodegradable products are ongoing in a few African universities, such as the University of Ghana. Countries such as Tanzania, Nigeria, Ghana, and Egypt have adopted policies that encourage the development and implementation of national recycling. For example, Nigeria uses waste materials as aggregates to produce construction materials such as tiles and bricks, tiles (Onu et al., 2018). In Tanzania, the involvement of the private sector under arrangements and local communities in solid waste management activities has created employment and job opportunities for a substantial number of jobless city residents, many of whom were previously unemployed women and youths. Solid waste management activities have been serving as a means of income generation for people taking part in the activities. Income generated is not only from wage payments but also from sales of items recovered from solid waste. Reused solid waste produced from items such as plastic bags, plastic containers, knives, spoons, and frying pans. Which are existing sales in major markets in the cities. In addition, the charges that service users pay "Refuse collection charges" are another good source of income for the Dar es Salaam City Council and the five Tanzania Municipalities (Mwesigye et al., 2009). Tanzania's Manners in which solid waste was previously managed have changed. For example, there are increasing signs of waste being segregated at source and to a large extent being stored in dust bins, sorting being done at communal waste collection points and, there being noticeable organized groups of people already involved in solid waste recycling. Increasingly, people's attitudes towards solid waste are gradually changing, so they no longer regard solid waste as a menace but rather as a commercial good that can be used to generate income and alleviate poverty (Mwesigye et al., 2009).

6.3 PROBLEMS

6.3.1 Ghana Management Practices Case Study

1.1 What are some of the key minerals produced in Ghana, and how significant is their contribution to the country's GDP?
1.2 How does the Ghanaian government demonstrate leadership in waste management within the mining sector?
1.3 What are some challenges faced in waste management within Ghana's mining industry?

6.3.2 Namibia's Mining Sector

2.1. What role does the mining sector play in the economy of Namibia?
2.2. How do mining training programs in Namibia contribute to skill development?
2.3 What are the main types of mines found in Namibia?
2.4 What measures are being taken to address industrial waste in Namibia?
2.5 What long-term goals exist regarding hazardous waste management in Namibia?
2.6 What challenges are associated with waste disposal sites in Namibia?
2.7 What can be inferred from Figure 6.5 regarding waste disposal sites in Namibia?

6.3.3 South African Management Practices

3.1 What are the primary sources of waste in South Africa's mining industry, and what percentage of waste is recycled?
3.2. How is fly ash utilized in South Africa, and what challenges does it present?
3.3. What innovative approaches are employed in South Africa for waste processing and resource recovery in the mining sector?

6.3.4 Botswana Management Practices

4.1. What are some of the key environmental challenges faced by Botswana as a result of its rapid development?
4.2. How does socioeconomic development contribute to the generation of solid waste in Botswana?
4.3. What strategies are proposed to minimize solid waste ending up in landfills in Botswana?
4.4. Describe the role of Collect-A-Can and Somarelang Tikologo in waste management in Botswana.
4.5. What is the primary method of waste disposal in Gaborone, Botswana, and how is it managed?

6.3.5 Nigeria Management Practices

5.1. What are some of the factors contributing to the increase in waste production in Nigeria?

5.2. How does waste disposal in Nigeria differ from standard waste management practices, and what are the consequences of these methods?
5.3. What is the current state of waste processing and resource recovery in Nigeria, and what are the challenges faced in implementing recycling programs?
5.4. How does the concept of circular economy apply to waste management in Nigeria, and what initiatives are being undertaken to promote sustainability?
5.5. What are some examples of sustainable waste management initiatives in other African countries, and how do they compare to practices in Nigeria?

REFERENCES

Adler, R. A., Claassen, M., Godfrey, L., & Turton, A. R. (2007). Water, mining, and waste: An historical and economic perspective on conflict management in South Africa. *The Economics of Peace and Security Journal*, *2*(2).

Assfour, B., Leoni, S., Yurchenko, S., & Seifert, G. (2011). Hydrogen storage in zeolite imidazolate frameworks: A multiscale theoretical investigation. *International Journal of Hydrogen Energy*, *36*(10), 6005–6013.

Bian, Z., Miao, X., Lei, S., Chen, S. E., Wang, W., & Struthers, S. (2012). The challenges of reusing mining and mineral-processing wastes. *Science*, *337*(6095), 702–703.

Botswana Selected Environmental Indicators Digest 2020. (n.d.). https://www.statsbots.org.bw/sites/default/files/publications/Botswana%20Selected%20Environmental%20Indicators%20Digest%202020.pdf

DEFF. (2020). National waste management strategy 2020. *Department of Environment Foretry and Fisheries – Republic of South African* (p. 71). https://www.dffe.gov.za/sites/default/files/docs/nationalwaste_management_strategy.pdf

Doka, G. (2008). *Life cycle inventory data of mining waste: Emissions from sulfidic tailings disposal*. Doka Life Cycle Assessments.

Draft, F. (2017). *Republic of Namibia ministry of environment and tourism national solid waste management strategy for Namibia*. https://doka.ch/SulfidicTailingsDisposalDoka.pdf

Festin, E. S., Tigabu, M., Chileshe, M. N., Syampungani, S., & Odén, P. C. (2019). Progresses in restoration of post-mining landscape in Africa. *Journal of Forestry Research*, *30*(2), 381–396.

Gericke, M., Neale, J. W., & Määttä, P. (2022). Biomining in Finland: Commercial application of heap and tank bioleaching technologies for nickel recovery. In Johnson, D. B., Bryan, C. G., Schlömann, M., Roberto, F. F. (Eds.). *Biomining technologies: Extracting and recovering metals from ores and wastes* (pp. 209–228). Springer International Publishing.

Hansen, K. (2015). *Overview of wastewater treatment in South Africa* (p. 63). Association for Water and Rural Development (AWARD) Publication.

Hogland, W., Visvanathan, A., MARQUES, M. and Manandhar, D.R. (2006). Landfill in Asia–Improving sanitation of landfill sites. *The waste management world* (pp. 87–96).

IFC. (n.d.). *EHS guidelines: Oil, gas, and mining*. https://www.ifc.org/en/insights-reports/2000/ehs-guidelines-oil-gas-mining

International Institute for Sustainable Development. (n.d.). *IGF case study: Mine waste management*. https://www.iisd.org/publications/brief/igf-case-study-mine-waste-management

Kamunda, C., Mathuthu, M., & Madhuku, M. (2016). An assessment of radiological hazards from gold mine tailings in the province of Gauteng in South Africa. *International Journal of Environmental Research and Public Health*, *13*(1), 138.

Mahla, D. K., Bhandari, S., Rahaman, M., & Khastgir, D. (2013). Morphology and cyclic voltammetry analysis of in situ polymerized polyaniline/graphene composites. *Journal of Electrochemical Science and Engineering*, *3*(4), 157–166.

Matsumoto, S., Shimada, H., & Sasaoka, T. (2016). The key factor of acid mine drainage (AMD) in the history of the contribution of mining industry to the prosperity of the United States and South Africa: A review. *Natural Resources*, *7*(7), 445–460.

Nwaila, G. T., Ghorbani, Y., Zhang, S. E., Frimmel, H. E., Tolmay, L. C., Rose, D. H., Nwaila, P. C., & Bourdeau, J. E. (2021). Valorisation of mine waste-part I: Characteristics of, and sampling methodology for, consolidated mineralised tailings by using Witwatersrand gold mines (South Africa) as an example. *Journal of Environmental Management, 295*, 113013. https://doi.org/10.1016/j.jenvman.2021.113013

Onipede, A. I. M., & Bolaji, B. O. (2004). Management and disposal of industrial wastes in Nigeria. *Nigerian Journal of Mechanical Engineering*, *2*(1), 49–58.

Onu, B., Pric, T., Surendran, S.S. and Ebie, S. (2018). Solid waste management: A critique of Nigeria's waste management policy. *The International Journal of Knowledge, Culture, & Change Management, 11*. Available at: [http://www.management-journal.com] (Accessed: 8 April 2024).

Ortega, F., Versino, F., López, O. V., & García, M. A. (2022). Biobased composites from agro-industrial wastes and by-products. *Emergent Materials*, *5*(3), 873–921.

Peša, I. (2022). Mining, waste and environmental thought on the Central African Copperbelt, 1950–2000. *Environment and History*, *28*(2), 259–284.

Petrik, L. F., White, R. A., Klink, M. J., Somerset, V. S., Burgers, C. L., & Fey, M. V. (2003, October). *Utilization of South African fly ash to treat acid coal mine drainage, and production of high quality zeolites from the residual solids* (Vol. 2022). Proceedings of the Ash Utilization Symposium, Lexington, KY.

Qadir, M., Wichelns, D., Raschid-Sally, L., McCornick, P. G., Drechsel, P., Bahri, A., & Minhas, P. S. (2010). The challenges of wastewater irrigation in developing countries. *Agricultural Water Management*, *97*(4), 561–568.

Schoeman, Y., Oberholster, P., & Somerset, V. (2021). A decision-support framework for industrial waste management in the iron and steel industry: A case study in Southern Africa. *Case Studies in Chemical and Environmental Engineering*, *3*, 100097.

Signé, L., & Johnson, C. (2021). *Africa's mining potential: Trends, opportunities, challenges and strategies* (Policy Brief 21/10). Policy Center for the New South.

Silva, A., Rosano, M., Stocker, L., & Gorissen, L. (2017). From waste to sustainable materials management: Three case studies of the transition journey. *Waste Management*, *61*, 547–557.

Suresh, S., & Vijayakumar, V. (2012). *Waste management in Botswana*. https://www.diva-portal.org/smash/get/diva2:500870/FULLTEXT01.pdf

Western Cape Government (2018). *Annual State of Waste Management Report 2018*. Available at: [https://d7.westerncape.gov.za/eadp/sites/eadp.westerncape.gov.za/files/atoms/files/Annual%20State%20of%20Waste%20Management%20Report%202018.pdf] (Accessed: 8 April 2025).

World Bank (2023). *Trends in Solid Waste Management*. Available at: https://datatopics.worldbank.org/what-a-waste/trends_in_solid_waste_management.html (Accessed: 8 April 2025).

Zvarivadza, T. (2018). Sustainability in the mining industry: An evaluation of the National Planning Commission's diagnostic overview. *Resources Policy*, *56*, 70–77.

7 Agricultural Waste

Thabo Falayi

7.1 INTRODUCTION

Agricultural waste represents one of the wastes found in Africa, more so where secondary value addition is minimal. However, the waste can also be viewed as a secondary resource and if utilized well, would result in a circular sustainable economy. Cassava waste (CW) is one of these and this chapter looks at how this waste can be used as a secondary resource. Cassava is known as a poor man's crop (Oghenejoboh et al., 2021). Nigeria produces 59.5% of the cassava in Africa. The source of CW is from the processing of the cassava for food by small-scale farmers (Malandula Chipeta et al., 2013). CW includes stems, peels, pulp, and wastewater. It is important to note that leaves and peels constitute the bulk of the waste (Moshi & Nges, 2017).

7.2 MANAGEMENT PRACTICES: CASSAVA PEELS (AFRICA)

Most cassava waste management practices are still at an experimental stage because there is no formalized high-scale industry to produce cassava products in Africa. The management practices discussed in this chapter are the best management practices that can be adopted by cassava-producing countries.

7.2.1 Waste Collection

The largest portion of agricultural waste in Nigeria has been shown to be CW (Oghenejoboh et al., 2021). Most of the waste from the cassava is collected from the processing sites (Achi et al., 2018) and this can be small-scale producers of a variety of different households. The collection is mostly not for reprocessing but as a means of waste disposal.

7.2.2 Waste Processing

The processing of CP includes washing with copious amounts of water to remove mud and debris (Achi et al., 2018). Though CW is disposed as is, it is recommended to boil the wastewater or the solid residue at about 100°C to destroy the cassava enzymes, which produce hydrogen cyanide through hydrolysis of linamarin and lotanstration (Ero & Okponmwense, 1978). Ensiling of CW is one way of processing the waste (Ubalua, 2007), where the CW is ground and immediately followed by salt addition to allow for anaerobic fermentation. The product of ensiling can be used as stock feed.

DOI: 10.1201/9781003377856-7

7.2.3 Resource Recovery

There are numerous resource recovery routes that have been postulated for CW but there are a few which are being implemented. These routes include production of energy through anaerobic digestion and production of ethanol through fermentation (Moshi & Nges, 2017).

7.2.4 Waste Disposal

CW is usually disposed indiscriminately (Oghenejoboh et al., 2021). This can be due to the fact that most processors of Cassava are peasant farmers who may not have the means or technical know-how to dispose of the CW. This is even made worse by dysfunctional municipalities in Africa (Muheirwe et al., 2022). CW is usually dumped in any nearby open sites or water bodies (Achi et al., 2018; Olukanni & Olatunji, 2018). The CW can also be used for landfilling purposes or can be burnt to provide an energy source (Achi et al., 2018). The resulting ash can then be disposed of in the vicinity of where the CW was being used as an energy source. At most 25% of the cassava peels can be used as livestock feed, this presents another avenue for CW disposal.

7.3 OPPORTUNITIES TO SHIFT TOWARDS A CIRCULAR ECONOMY

There are various ways to incorporate circular economy concepts into the management of CW. These include fermentation of CW leading to reduction of cyanide. The fermented CW can then be used as stock feed, production of adsorbents for water treatment, and production of ethanol (Ubalua, 2007). CW can also be used as organic compost and for production of biofuels (Nizzy & Kannan, 2022). Cassava peels (CP) constitute 10–20% of the tuber weight, and with about an annual production of 140 million tonnes of cassava tuber, there is enough supply of this waste material to be used as a bio-sorbent (Kayiwa et al., 2021). This section therefore focuses on the use of CP as an adsorbent for metal ions in wastewater. The section also incorporates the use of machine learning in the prediction of performance of CP as an adsorbent for various wastewater to improve the applicability.

CP has been used as an adsorbent for metal ions and dyes (Rajeshwarisivaraj et al., 2001) where the removal rate was above 99% for dyes like malachite green and methylene blue. Adsorption involves sequestration of metal ions onto synthetic or natural materials (Nadeem et al., 2014). Adsorption is one of the most efficient and cost-effective for the removal of metal ions from wastewater (Sowmya et al., 2023). Adsorption is also seen as highly sensitive to metal ion pollutants, thereby increasing the metal ion removal efficiency (Sharma et al., 2022). CP has also been used as a biosorbent for Ni^{2+} metal ions, achieving a maximum adsorption capacity of 57 mg/g at a pH of 4.5 (Kurniawan et al., 2011). The versatility of CP as a biosorbent has also been demonstrated in its use in the removal of chlorpyrifos from wastewater (Schwantes et al., 2022). Activation of CP has also been employed to improve the performance of the biosorbent, resulting in its use in the removal of naphthol blue-black,

Cd, Cr, and Ni (Alongamo et al., 2021; Astuti et al., 2020; Poolprasert & Chorchong, 2020). The activation includes the use of KOH, H_3PO_4, and $ZnCl_2$ together with calcination. It is important to note however that most of the research has not incorporated machine learning in order to test the applicability of CP peels as bio sorbent for a number of effluents.

7.4 ISOTHERMS

The Freundlich isotherm is one of the adsorption empirical isotherms (Wang & Guo, 2020). The non-linear equation for the Freundlich is given as

$$q_e = K_F C_e^{\frac{1}{n}} \qquad \text{(eqn. 7.1)}$$

where q_e (mg/g) and C_e (mg/) are adsorbed amounts per mass of adsorbent and adsorbate concentrations at equilibrium, K_F ($L^{1/n}.mg^{1-1/n}.g^{-1}$) and n are constants. The use of non-linear isotherms reduces errors associated with linearization which can be as high as 40% (Wang & Guo, 2020). The Freundlich isotherms represent multilayer adsorption on heterogeneous adsorbents.

The Langmuir isotherm is one of the chemical adsorption isotherms (Wang & Guo, 2020). The nonlinear equation for the Langmuir equation is given as

$$q_e = \frac{q_m K_L C_e}{1 + K_L C_e} \qquad \text{(eqn. 7.2)}$$

where K_L (L/m) is the ratio of the adsorption rate and desorption rate, qm (mg/g) is the maximum adsorption capacity of the adsorbent. The Langmuir isotherm is based on the premise of monolayer adsorption, homogeneous distribution of adsorption, constant adsorption energy, and the interaction between adsorbate molecules is negligible.

Sips Model

This is a hybrid Langmuir and Freundlich isotherm model. The non-linear equation is given as

$$q_e = \frac{q_{ms} K_s C_e^{ns}}{1 + K_s C_e^{ns}} \qquad \text{(eqn. 7.3)}$$

where q_{ms} (mg/g) is the maximum adsorbed amount, Ks (L^{ns}/mg^{ns}) and n_s is the Sips constant. This is a three-parameter isotherm that can be used for both homogeneous and heterogeneous adsorption. The greater from unity n_s, the more heterogeneous the adsorption system.

7.5 ADSORPTION KINETICS

The pseudo first-order kinetic (PFO) is an empirical kinetic equation (Wang & Guo, 2020). The non-linear equation for PFO is given as

$$q_t = q_e\left(1 - e^{-k_1 t}\right) \qquad \text{(eqn. 7.4)}$$

where q_t is the amount of adsorbate adsorbed per mass of adsorbent after a time *t*. The PFO is associated with instances where external/internal diffusion is the rate-controlling step thereby relating to physisorption (Wang & Guo, 2020).

The pseudo second-order kinetic (PSO) is an empirical kinetic equation. The non-linear equation for PSO is given as

$$q_t = \frac{q_e^2 k_2 t}{1 + q_e k_2 t} \qquad \text{(eqn.7.5)}$$

The PSO is mostly dominated by adsorption at the adsorption site is mostly chemisorption (Zhang, 2019).

7.6 ARTIFICIAL NEURAL NETWORK

The artificial neural network (ANN) algorithm is said to mimic the way the brain functions (Fertu et al., 2022). The ANN is made up of linked neurons which are connected via weights (Hafsa et al., 2020). The most common ANN is the feed-forward multilayer perceptron which has at least an input and output node. The training of the ANN uses backward propagation algorithms (Kennedy et al., 2015). Each node is linked to each node in the next layer, leading to the prediction of outputs from inputs through nonlinear functions. There are many advantages of using ANN and these include that (i) the algorithm is learnt from the data provided without the need to understand the science governing the process (ii) ANN is resilient to noise (Kennedy et al., 2015).

7.7 MATERIALS

CP was collected from a local market in Blantyre, Malawi. They were washed to remove debris and dried at 105°C till the dried CP had a constant weight. The dried CP was crushed using mortar and pestle till all the ground CP passed the 75 μm sieve. Analytical grade $FeSO_4.7H_2O$, $NiSO_4.5H_2O$, and $CuSO_4.5H_2O$ were used to synthesize the artificial wastewater. Analytical grade HCl was used for adjusting the pH of the synthesized wastewater. All chemicals were supplied by Sigma Aldrich South Africa.

7.8 METHOD

7.8.1 XRF

The CP sample was pelletized into a pellet, with subsequent pellet analysis using a Rigaku (NEX CG) X-ray fluorescence machine (XRF).

7.8.2 External Surface Area

The external surface area was determined using a laser diffraction technique (Mastersizer, 2000) by transforming the number distribution using second moments.

7.8.3 Point of Zero Charge

The point of zero-charge (pHpzc) of CP was determined by adding CP to respective 100 ml solutions of 0.01 M KNO_3 solutions. The pH of the solutions ranged from 2 to 12. The mixtures were agitated for 24 hours. At the end of 24 hours, the final pH was measured. The pH difference was then plotted versus the initial pH. pHpzc was determined as the point the graph cuts the x-axis (El Alouani et al., 2023).

7.8.4 Effect of Solid Loading

A 50 ppm (Fe(II), Cu(II), and Ni(II)) wastewater was prepared by dissolving the appropriate amounts of the respective metal salts. The pH of the wastewater was adjusted to 1 using a drop-wise addition of 1 M HCl. A 2–10% mass/volume (m/v) loading of the CP adsorbent was added to the respective 100 ml of wastewater in a 250 ml conical flask at 25°C in a thermoshaker. The mixtures were agitated for 2 hours at 200 rpm. At the end of 2 hours, the mixtures were filtered and the filtrates were subjected to metal ion analysis using an atomic absorption spectrometer.

7.8.5 Effect of Initial pH

The initial metal ion concentration was maintained at 50 ppm. The pH was adjusted to 2, 3, 4, and 5, respectively. For each pH level, 100 ml of wastewater was mixed with 6 g of CP and agitated at 200 rpm for 2 hours at 25°C in a thermoshaker. At the end of 2 hours, the mixtures were filtered and the filtrates were subjected to metal ion analysis using an atomic absorption spectrometer.

7.8.6 Effect of Initial Concentration

Respective wastewater was prepared at 100, 200, 400, 500, 600, 700, 800, 900 and 1000 ppm (Fe(II), Cu (II), and Ni(II)) concentrations. The pH was adjusted to 4 and the solid loading was maintained at 6% m/v. The mixtures were agitated for 2 hours at 200 rpm. At the end of 2 hours, at 25°C, the mixtures were filtered and the filtrates were subjected to metal ion analysis using an atomic absorption spectrometer.

7.8.7 Effect of Time and Temperature

A 500 ppm wastewater (pH 4) was prepared and mixed with CP at 6% m/v solid loading. The mixtures were agitated at 200 rpm at 25°C for 20, 40, 60, 80, 100 and 120 min, respectively. At the end of each respective time, the mixtures were filtered and the filtrates were subjected to metal ion analysis using an atomic absorption spectrometer. The above experiment was also repeated at 35°C and 45°C.

7.8.8 Machine Learning

MATLAB R2021a was used for the ANN machine learning using all the raw data for the prediction of CP performance as a biosorbent.

7.8.9 Statistical Analysis

All experiments were done in triplicate with the reported result being the average of samples which were within 10% of each other. Validation of the ANN can be done through mean square error (MSE) and R^2 correlation coefficient (Abbasi et al., 2022). The formulas for R^2 and MSE are shown in equations 7.6 and 7.7, respectively.

$$R^2 = 1 - \frac{\sum_{i=1}^{N}\left(x_i - x_i^{ANN}\right)^2}{\sum_{i=1}^{N}\left(x_i - \underline{x}\right)^2} \quad \text{(eqn. 7.6)}$$

$$MSE = \frac{\sum_{i=1}^{N}\left(x_i - x_i^{ANN}\right)^2}{N} \quad \text{(eqn 7.7)}$$

7.9 RESULTS AND DISCUSSION

7.9.1 XRF Analysis

Table 7.1 shows the XRF analysis of CP.

The XRF analysis of CP was dominated by CaO and K_2O. These base oxides are usually responsible for pH change during adsorption, leading to increased removal of metal ions (Sithole et al., 2023). The CP also has a very high loss on ignition showing its biological nature.

7.9.2 Surface Area

Laser diffraction analysis of the <75 μm particles showed a Sauter mean of 11.5 μm, a DeBroukere mean of 39.9 μm with an external surface area of 824.12 m^2/g, which was very comparable to a number of alkaline activated adsorbents (Abegunde et al., 2020).

TABLE 7.1
XRF Analysis of CP

Constituent	% mass/mass
CaO	3.89
Fe_2O_3	0.54
SiO_2	0.80
Al_2O_3	0.95
SO_3	1.01
P_2O_5	0.15
K_2O	4.58
LOI	87.83

7.9.3 pH of Zero Charge

The pH of zero charge (pHpzc) was found to be 3.9 (Figure 7.1). It therefore means that at pH<3.9, the surface of the CP is positively charged, while at pH>3.9 the surface of CP is negatively charged (El Alouani et al., 2023).

7.9.4 Effect of Solid Loading

The variation in metal removal with solid loading as shown in Figure 7.2. An increase in metal ion removal with solid loading was observed (Figure 7.2). This was mainly

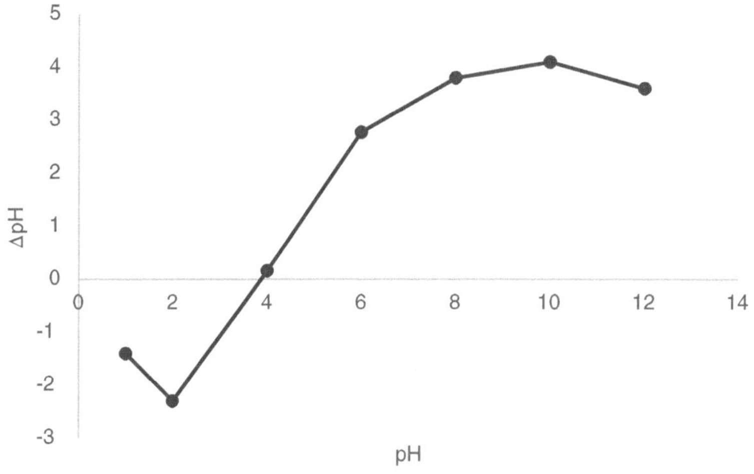

FIGURE 7.1 CP pHpzc plot.

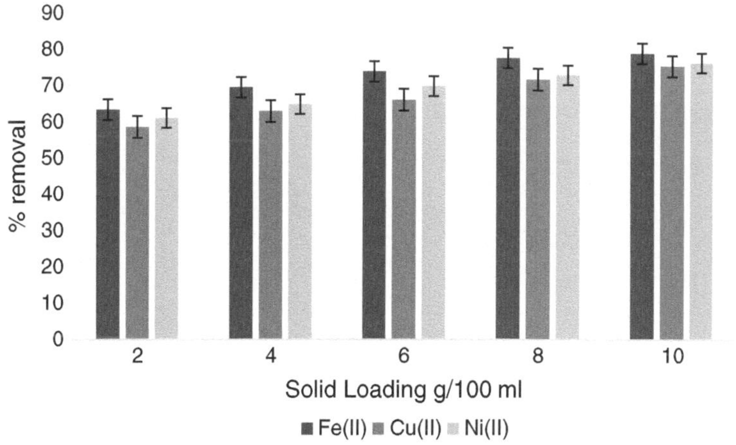

FIGURE 7.2 Variation in metal ion removal with solid loading.

due to the increase in the available adsorption sites with an increase in solid loading. The increase in metal ion removal from 6% to 10% was not statistically significant at the calculated p-value was 0.216. This may have been due to competitive adsorption since the wastewater was very acidic (pH 1) with a very high concentration of H^+ ions. A solid loading of 6% was used for subsequent experiments.

7.9.5 Effect of Initial pH

pH is one of the most important factors in adsorption and the highest removal was at pH 4 and 5 (Figure 7.3). This agrees well with the pHpzc experiment (Figure 7.1). pH 4 is just above the pHpzc value of 3.9. There was no statistical significance in metal ion removal between pH 4 and 5 maybe due to metal ion exhaustion (p value 0.498). pH 4 was then used for all subsequent experiments.

7.9.6 Effect of Initial Concentration

There was a decrease in metal ion removal with an increase in initial metal ion concentration (Figure 7.4). This may have been due to an increase in competitive adsorption and adsorption site depletion with an increase in metal ion concentration at a fixed solid loading. The decrease in metal ion removal was significant from 500 – 1000 ppm (p-value 0.01). Initial concentration of 500 ppm was used for the rest of the experiment as the residual metal ion concentration after adsorption was within the statutory limits.

7.9.7 Effect of Time

There was an increase in adsorption capacity with time meaning more was adsorbed with an increase in time (Figure 7.5). The equilibrium for all metals was reached

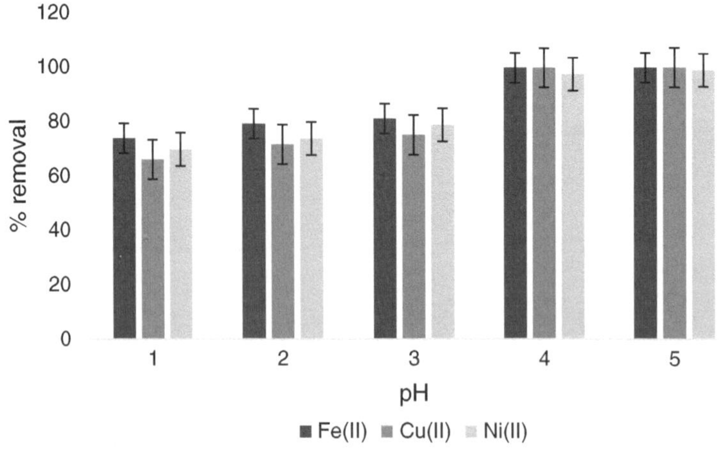

FIGURE 7.3 The variation of metal ion removal with initial pH.

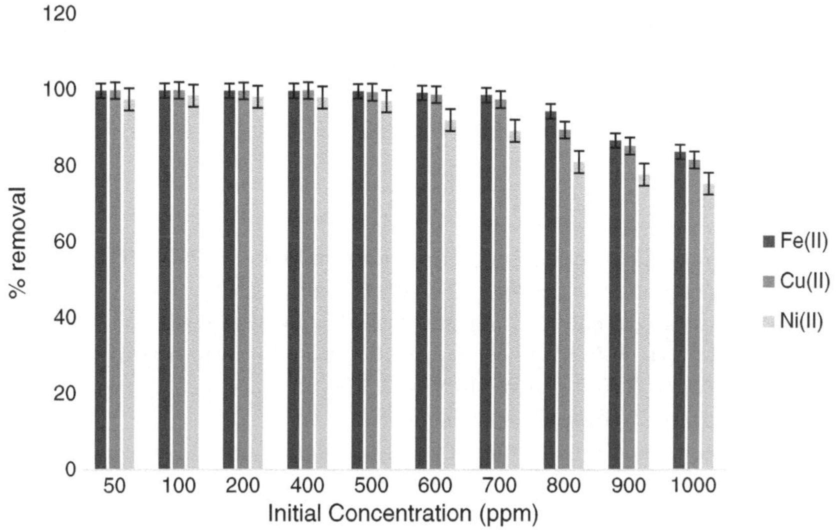

FIGURE 7.4 Variation in metal removal with initial metal ion concentration.

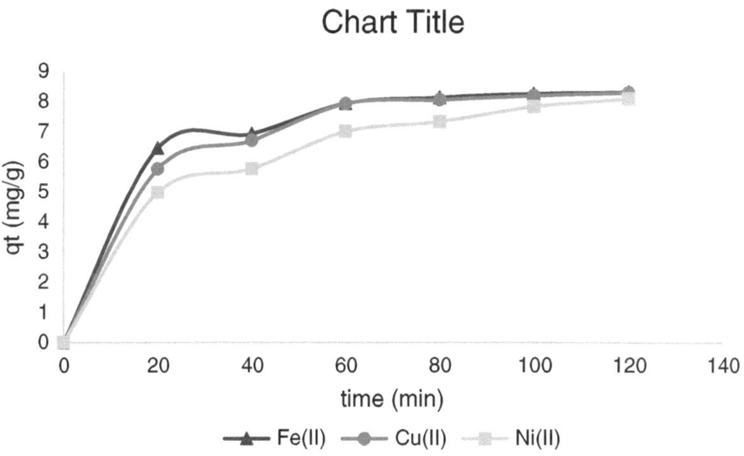

FIGURE 7.5 Variation adsorption capacity with time.

around 100 minutes as the statistical difference between adsorption capacity and 100 and 120 min was insignificant (p-value 0.42). The adsorption proceeds at a fast rate during the first 60 min due to available adsorption sites, and equilibrium is reached due to either metal ion depletion or adsorption site exhaustion (Choudhry et al., 2021).

7.9.8 Effect of Temperature

There was an increase in q_e with an increase in temperature (Figure 7.6). This was due to energy provided which increased the interaction between adsorbent and adsorbate. The order of adsorption was $Fe^{2+}>Cu^{2+}>Ni^{2+}$.

7.9.9 Adsorption Isotherms

The Sip and Freundlich isotherm plots are shown in Figures 7.7–7.8. The Langmuir isotherm was not plotted as the correlation coefficient was below 0.6.

The use of linearized equations for isotherms was equally avoided because linearization errors of up to 40% have been reported. The nonlinear isotherms were solved using the MATLAB R2021a curve fitting tool. The Sips model could be used to best describe the adsorption of metal ions onto CP.

The Sips model had a better fit, exhibiting a higher correlation coefficient and the lower sum of squared estimate of errors (SSE) (Table 7.2). The value for the Sips coefficient was greater than 1 showing that the adsorption was heterogeneous. The Sips model can also be used for systems at high concentrations, which fits well with the experimental data (Foo & Hameed, 2010).

7.9.10 Kinetics

The PFO and PSO plots are shown in Figures 7.9–7.10.

The adsorption of Fe^{2+}, Cu^{2+}, and Ni^{2+} onto CP could be modelled well using PFO as the correlation coefficient was higher than 0.97 (Table 7.3). The difference between experimental and calculated q_e was below 0.3%. The PSO could not model the adsorption as it had a lower correlation coefficient, higher values for

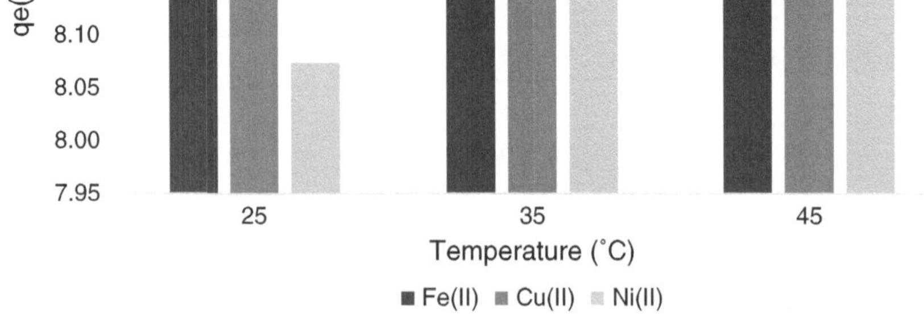

FIGURE 7.6 Variation in q_e with temperature.

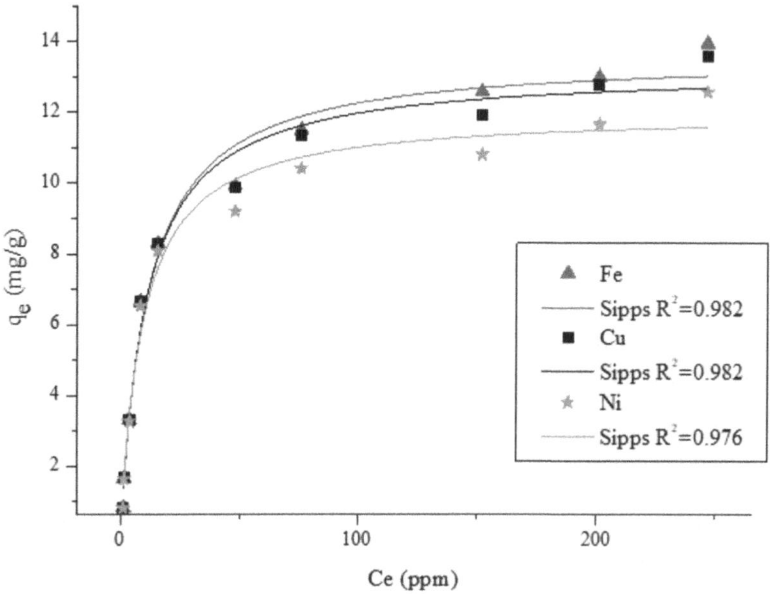

FIGURE 7.7 Sips Model.

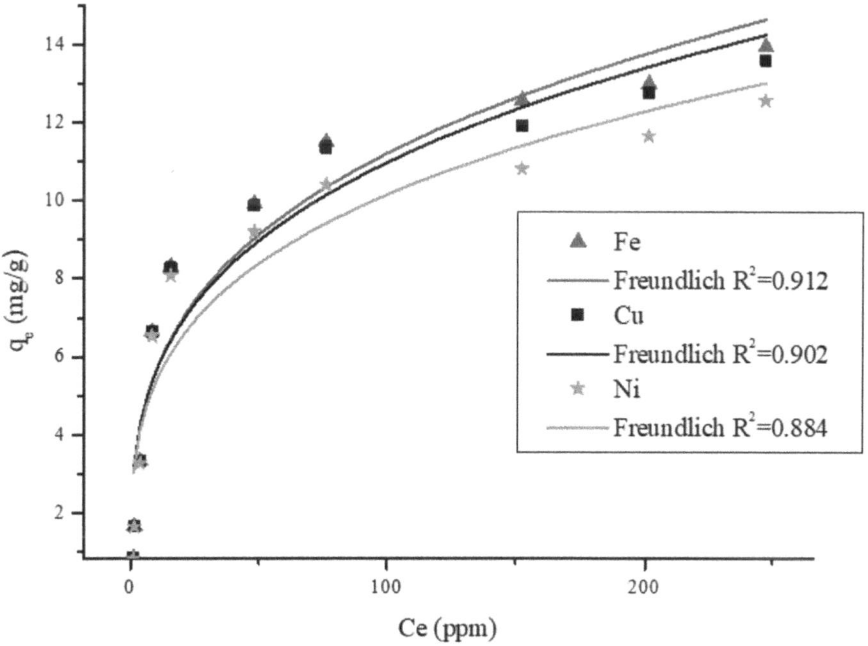

FIGURE 7.8 Freundlich isotherm.

TABLE 7.2
Non-linear Isotherm Parameters

Metal ion	Fe^{2+}	Cu^{2+}	Ni^{2+}
Sips Model			
Qms (mg/g)	12.054	12.045	11.82
Ks (L^{ns}/mg^{ns})	4.181	5.041	1.202
nc	4.545	4.916	5.657
SSE	0.064	0.156	0.145
R^2	0.982	0.982	0.976
Freundlich			
K_F ($L^{1/n}.mg^{1-1/n}.g^{-1}$)	2.876	2.900	2.855
n	3.385	3.461	3.631
SSE	0.1465	0.2119	0.2532
R^2	0.912	0.902	0.884

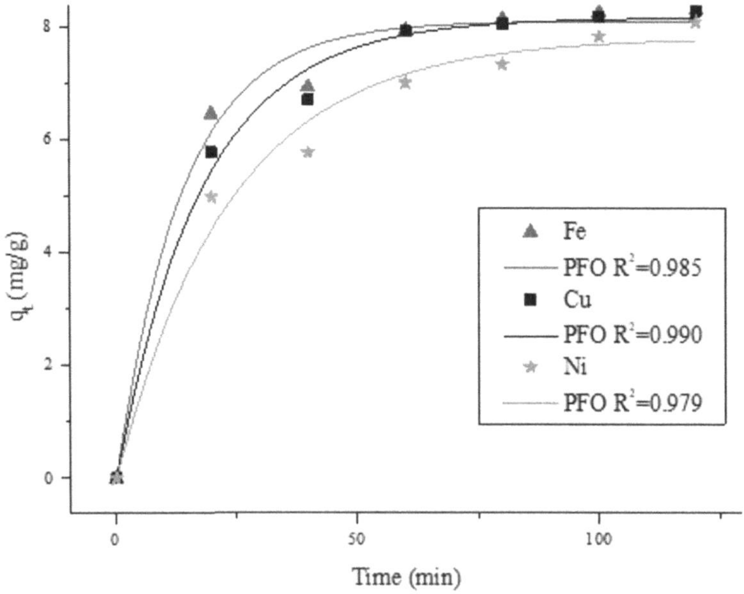

FIGURE 7.9 PFO kinetic plot.

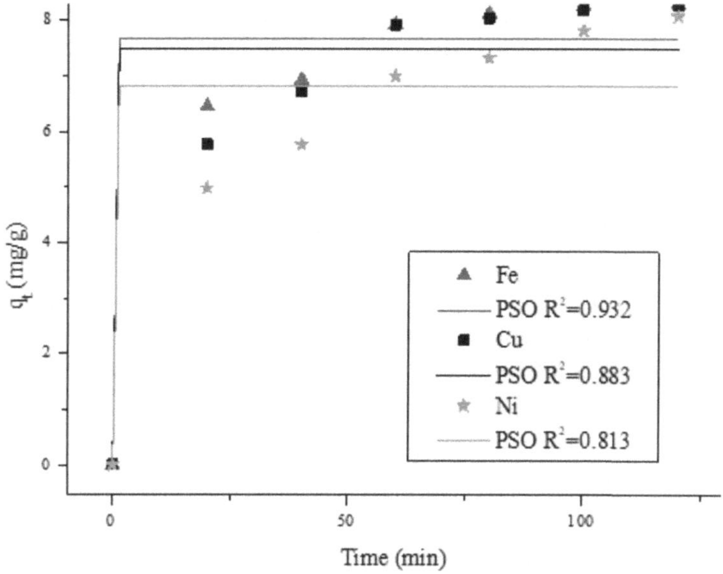

FIGURE 7.10 PSO isotherm.

TABLE 7.3
Kinetic Parameters

	PFO		
Metal ion	**Fe(II)**	**Cu(II)**	**Ni(II)**
Qe calculated (mg/g)	8.09	8.151	7.801
Qe experimental (mg/g)	8.30	8.27	8.07
K_1 (mol $L^{-1})^{1-n}$ s^{-1}	0.071	0.055	0.041
SSE	0.656	0.429	0.832
R^2	0.986	0.990	0.979
PSO			
Qe calculated (mg/g)	7.661	7.479	6.825
Qe experimental (mg/g)	8.30	8.27	8.07
K_2 (g mg^{-1} min^{-1})	–2.777	–1.534	2.10
SSE	3.042	5.201	7.362
R^2	0.932	0.883	0.813

SSE, a $\Delta q_e\%$ of at least 0.6% and negative kinetic constants. The PFO can be used to model physisorption which agrees well with the thermodynamic calculations (Figure 7.11).

7.9.11 Thermodynamics

A plot of ΔG versus T yielded linear graphs from which the thermodynamic parameters for the adsorption of Fe(II), Cu(II), and Ni(II) (Figure 7.11). The adsorption of the respective metal ions onto CP was physisorption since the calculated ΔG was less than -20 kJ/mol. The adsorption of the metals was feasible due to a positive entropy value for all metal ions. An increase in temperature resulted in more negative ΔG values (Table 7.4), meaning the adsorption became more feasible at higher temperatures, further supporting that the process was endothermic.

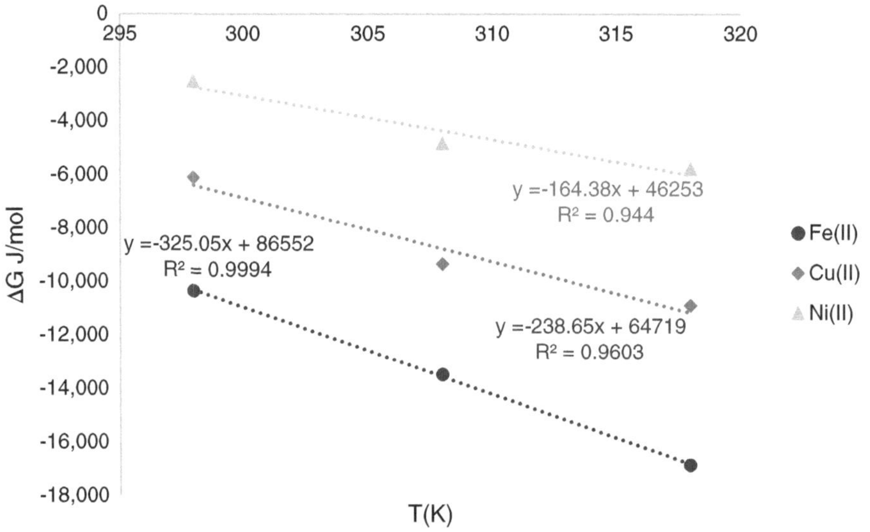

FIGURE 7.11 Thermodynamic plot.

TABLE 7.4
Thermodynamic Parameters

	ΔG @298K (kJ/mol)	ΔG @308K (kJ/mol)	ΔG @318K kJ/mol	ΔH (kJ/mol)	ΔS (J/mol)
Fe^{2+}	−10.356	−13.475	−16.857	86.552	325
Cu^{2+}	−6.120	−9.347	−10.892	64.719	239
Ni^{2+}	−2.502	−4.840	−5.790	46.253	164

7.9.12 Regeneration

Figure 7.12 shows the adsorption cycles for CP.

There was a decrease in metal removal with an increase in reuse cycles (Figure 7.12). The reduction in metal ion removal was not significant for the first two cycles; however, there was a 4.8%, 6.6%, and 8.4% drop in metal ion removal from the fourth cycle to the fifth cycle for Fe(II), Cu(II), and Ni(II), respectively. This may have been caused by the denaturing of the CP with each regeneration cycle as 0.005 M HCl was used as a regeneration agent. It is however important to note that water could not be used as a regeneration agent as the metal ion removal in the second cycle was below 30%, while a higher concentration of HCl could not be used as the second cycle removal was below 35%, probably due to the denaturing effect of the acid. HCl was chosen as it has been used on other adsorbents (Falayi & Ikotun, 2021).

7.9.13 Machine Learning Using Artificial Neural Network

The Levenberg–Marquardt algorithm was found to be the best to predict the performance of CP as an adsorbent. The number of neurons was kept below 11 in order to avoid overfitting (Kennedy et al., 2015) as per the formula written below

$$n_h \leq 2n_i + 1 \qquad \text{(eqn 7.8)}$$

where n_h is the number of hidden nodes, and n_i is the number of inputs. Figure 7.13 shows the architecture of the ANN used. The ANN had 10 hidden neurons with 5 inputs and 3 outputs. The inputs were pH, initial metal ion concentration, adsorption time, adsorption temperature, and solid loading, while the respective residual metal ion concentrations were the outputs.

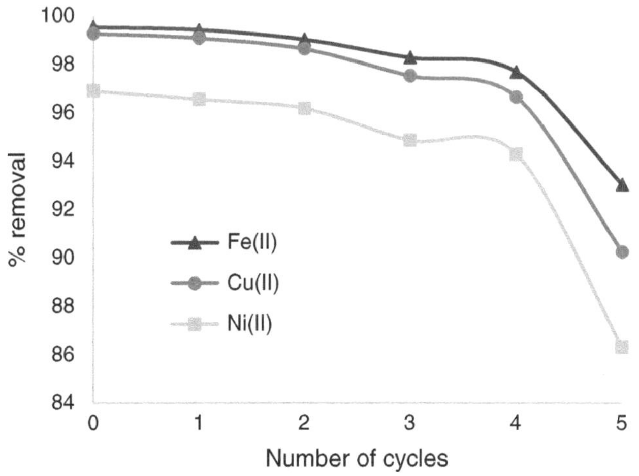

FIGURE 7.12 Variation in metal ion removal with regeneration.

7.9.14 Architecture of the ANN

7.9.14.1 Validation of ANN Performance

The best validation performance was at epoch 6 with an MSE of 7.7. The MSE for training and testing was 0.7 and 1, respectively (Figure 7.14). The lower the MSE, the more accurate the ANN.

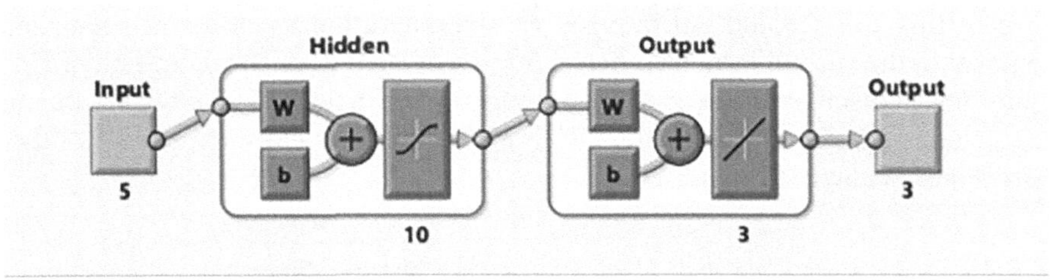

FIGURE 7.13 ANN architecture with 5 inputs and 10 hidden neurons and 3 output neurons and 3 outputs.

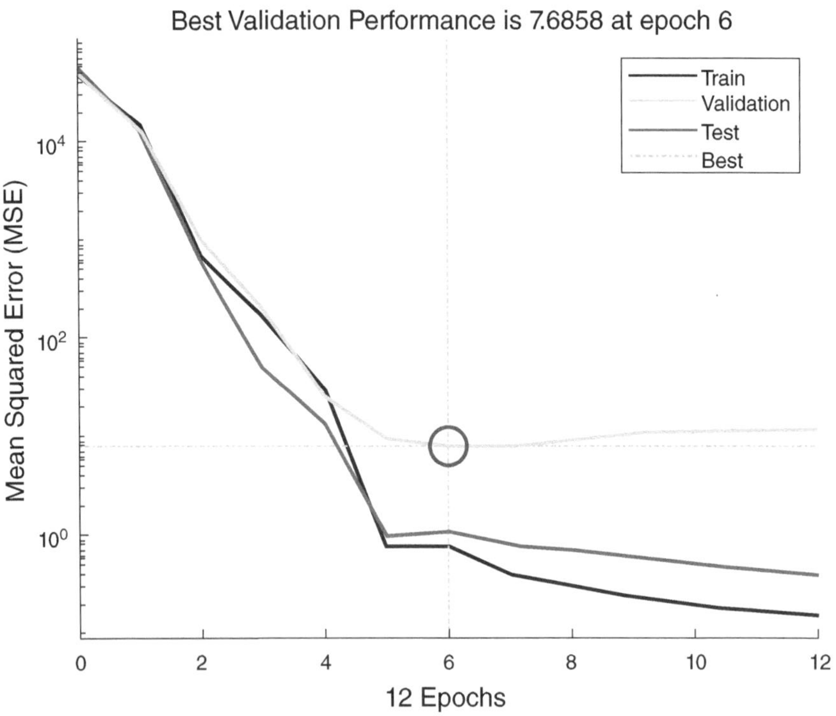

FIGURE 7.14 ANN validation performance.

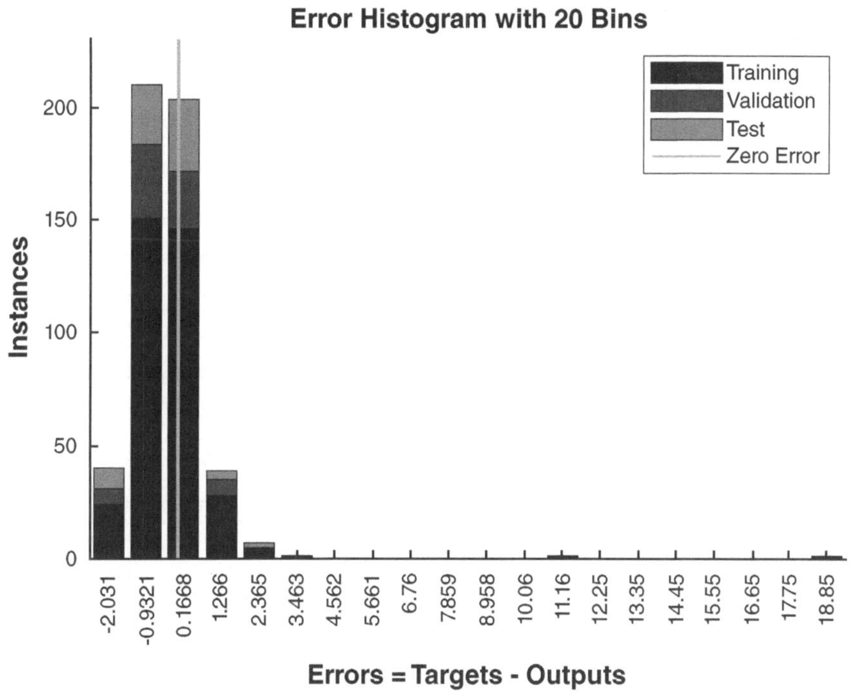

FIGURE 7.15 ANN error histogram.

Error histograms of the ANN showing the bulk of observations near the zero error line.

The bins close to the zero error line were the highest in terms of frequency, meaning that the model performance is good, as most errors are close to the zero error line (Figure 7.15). The error histogram can also be used to determine whether the extreme values are outliers or not (Younis & Kayi, 2019). The ANN model had an overall correlation coefficient of 0.97 (Figure 7.16) which showed that it could be used to adequately predict the performance of CP as an adsorbent for heavy metal ions.

7.9.15 Comparison

The CP used in this research was comparable with other studies with the recycle and ANN adding another layer to the body of knowledge (Table 7.5).

7.10 CONCLUSION

CP could be used as an adsorbent for Fe(II), Cu(II), and Ni(II) as a 6% solid loading of CP resulted in a 99.5%, 99.3%, and 96.9% removal of the respective metal ions.

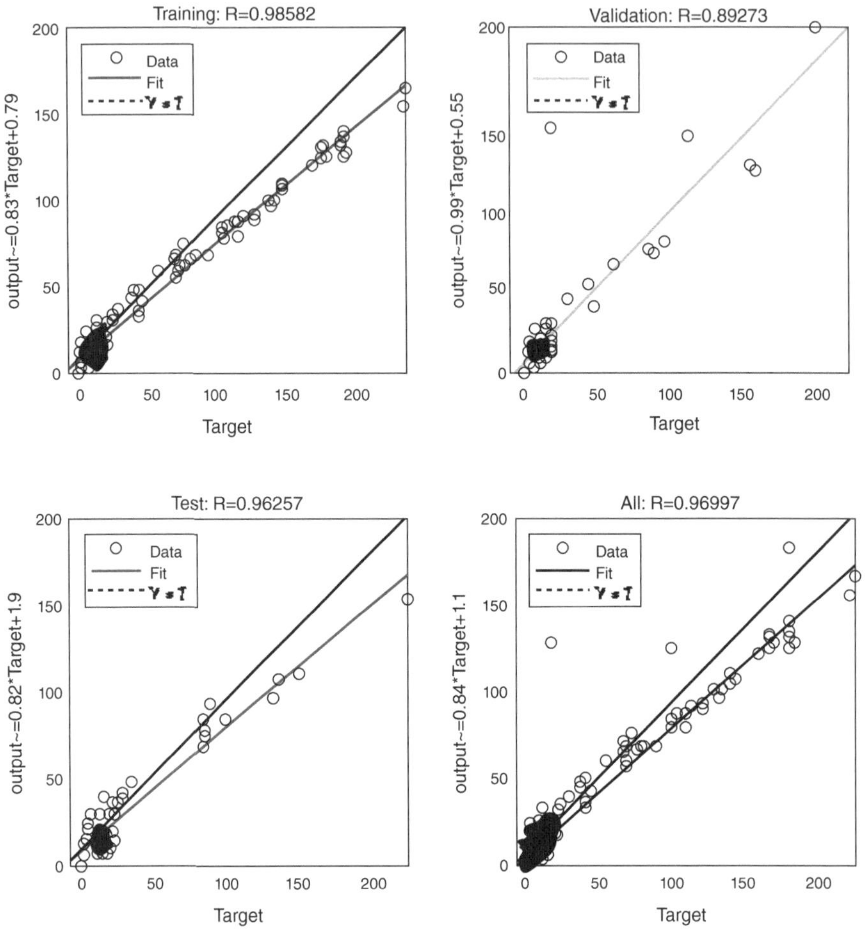

FIGURE 7.16 ANN performance.

The best conditions for metal ion adsorption were a solution pH of 4, initial metal ion concentration of 500 ppm, and adsorption time of 2 hours. The adsorption could be well modelled using the Sips isotherm and pseudo first-order kinetic model. The adsorption process was dominated by physisorption. The adsorption of metal ions onto CP was endothermic, characterized by positive entropy values. The CP could be generated using HCl and could be reused for four cycles. A 5,10,3 ANN architecture model could be used to predict the performance of the CP as an adsorbent for heavy metal ions. This study therefore provides an opportunity for the sustainable use of CP as an adsorbent for heavy metal ions.

TABLE 7.5
Comparison

Metal ion concentration	Maximum adsorption capacity	Regeneration	Machine learning	Reference
Cr(III), 110 ppm	2.68 mg/g	Not done	Not done	(Kristianingrum et al., 2022)
Cd(II) 5–200 ppm Cr(III) 5–200 ppm Pb(II) 5–200 ppm	19.54 mg/g 49.37 mg/g 42.46 mg/g	Not done	Not done	(Schwantes et al., 2016)
Pb(II) 5.30 ppm Cu(II) 4.0 ppm	5 mg/g 8 mg/g	Not done	Not done	(Owamah, 2014)
Fe(II) 50–1000 ppm Cu(II) 50–1000 ppm Ni(II) 50–1000 ppm	12.05 mg/g 12.05 mg/g 11.82 mg/g	4 cycles	8,10,3 ANN with R2 of 0.97	This study

REFERENCES

Abbasi, E., Alavi Moghaddam, M. R., & Kowsari, E. (2022). A systematic and critical review on development of machine learning based-ensemble models for prediction of adsorption process efficiency. *Journal of Cleaner Production*, *379*(P1), 134588. https://doi.org/10.1016/j.jclepro.2022.134588

Abegunde, S. M., Idowu, K. S., Adejuwon, O. M., & Adeyemi-Adejolu, T. (2020, August). A review on the influence of chemical modification on the performance of adsorbents. *Resources, Environment and Sustainability*, *1*, 100001. https://doi.org/10.1016/j.resenv.2020.100001

Achi, C. G., Coker, A. O., & Sridhar, M. (2018). Cassava processing wastes: Options and potentials for resource recovery in Nigeria. In S. K. Ghosh (Ed.), *Utilization and management of bioresources* (pp. 77–89). Springer Nature.

Alongamo, B. A. A., Ajifack, L. D., Ghogomu, J. N., Nsami, N. J., & Ketcha, J. M. (2021). Activated carbon from the peelings of cassava tubers (Manihot esculenta) for the removal of nickel(II) ions from aqueous solution. *Journal of Chemistry*, *2021*. https://doi.org/10.1155/2021/5545110

Astuti, W., Hidayah, M., Fitriana, L., Mahardhika, M. A., & Irchamsyah, E. F. (2020, June). Preparation of activated carbon from cassava peel by microwave-induced H3PO4 activation for naphthol blue-black removal. *AIP Conference Proceedings*, *2243*. https://doi.org/10.1063/5.0001464

Choudhry, A., Sharma, A., Khan, T. A., & Chaudhry, S. A. (2021, March). Flax seeds based magnetic hybrid nanocomposite: An advance and sustainable material for water cleansing. *Journal of Water Process Engineering*, *42*, 102150. https://doi.org/10.1016/j.jwpe.2021.102150

El Alouani, M., Alehyen, S., Aouan, B., Mabrouki, J., & Saufi, H. (2023). Application of mesoporous fly ash as a low-cost adsorbent for effective removal and immobilization of hazardous pollutants in aqueous environments. *Modeling Earth Systems and Environment*. https://doi.org/10.1007/s40808-023-01715-4

Ero, N. R., & Okponmwense, M. (1978). *The effect of cassava waste on the environment and its implications on the national economy*. Global Academic group.

Falayi, T., & Ikotun, B. D. (2021). A comparison between ferrochrome slag and gold mine tailings based geopolymers as adsorbents for heavy metals in aqueous solutions: Analyzing reusability and sustainability. *Korean Journal of Chemical Engineering*, *38*(4), 816–825. https://doi.org/10.1007/s11814-020-0731-y

Fertu, D. I., Dragoi, E. N., Bulgariu, L., Curteanu, S., & Gavrilescu, M. (2022). Modeling the biosorption process of heavy metal ions on soybean-based low-cost biosorbents using artificial neural networks. *Processes*, *10*(3), 1–24. https://doi.org/10.3390/pr10030603

Foo, K. Y., & Hameed, B. H. (2010). Insights into the modeling of adsorption isotherm systems. *Chemical Engineering Journal*, *156*(1), 2–10. https://doi.org/10.1016/j.cej.2009.09.013

Hafsa, N., Rushd, S., Al-Yaari, M., & Rahman, M. (2020). A generalized method for modeling the adsorption of heavy metals with machine learning algorithms. *Water*, *12*(12), 3490. https://doi.org/10.3390/w12123490

Kayiwa, R., Kasedde, H., Lubwama, M., & Kirabira, J. B. (2021, July). The potential for commercial scale production and application of activated carbon from cassava peels in Africa: A review. *Bioresource Technology Reports*, *15*, 100772. https://doi.org/10.1016/j.biteb.2021.100772

Kennedy, M. J., Gandomi, A. H., & Miller, C. M. (2015). Coagulation modeling using artificial neural networks to predict both turbidity and DOM-PARAFAC component removal. *Journal of Environmental Chemical Engineering*, *3*(4), 2829–2838. https://doi.org/10.1016/j.jece.2015.10.010

Kristianingrum, S., Sulistyani, S., & Larastuti, A. R. (2022). The effectiveness of active carbon adsorbent of cassava peel (Manihot Esculenta Cranzts) in reduce level of chromium metal in tannery liquid waste. *Indonesian Journal of Chemistry and Environment*, *5*(2), 58–67. https://journal.uny.ac.id/index.php/ijoce/article/view/18813

Kurniawan, A., Kosasih, A. N., Febrianto, J., Ju, Y. H., Sunarso, J., Indraswati, N., & Ismadji, S. (2011). Evaluation of cassava peel waste as low-cost biosorbent for Ni-sorption: Equilibrium, kinetics, thermodynamics and mechanism. *Chemical Engineering Journal*, *172*(1), 158–166. https://doi.org/10.1016/j.cej.2011.05.083

Malandula Chipeta, M., Bokosi, J. M., & Chipeta, M. M. (2013). Status of Cassava (Manihot esculenta) production and utilization in Malawi. *International Journal of Agronomy and Plant Production*, *4*(S), 3637–3644. http://www.ijappjournal.com

Moshi, A. P., & Nges, I. (2017). Sustainable management of cassava processing waste for promoting rural development. In C. KLEIN (Ed.), *Handbook on cassava production, potential uses and recent advances* (1st ed., pp. 129–148). Nova Publishers.

Muheirwe, F., Kombe, W., & Kihila, J. M. (2022). The paradox of solid waste management: A regulatory discourse from Sub-Saharan Africa. *Habitat International*, *119*(April 2021), 102491. https://doi.org/10.1016/j.habitatint.2021.102491

Nadeem, R., Nadeem, M., Afzal, A., & Asif, M. (2014). Potential of NaOH pretreated Mangifera indica waste biomass for the mitigation of Ni (II) and Co (II) from aqueous solutions. *Journal of the Taiwan Institute of Chemical Engineers*, *45*(3), 967–972. https://doi.org/10.1016/j.jtice.2013.09.012

Nizzy, A. M., & Kannan, S. (2022). A review on the conversion of cassava wastes into value-added products towards a sustainable environment. *Environmental Science and Pollution Research*, *29*(46), 69223–69240. https://doi.org/10.1007/s11356-022-22500-3

Oghenejoboh, K. M., Orugba, H. O., Oghenejoboh, U. M., & Agarry, S. E. (2021, February). Value added cassava waste management and environmental sustainability in Nigeria: A review. *Environmental Challenges*, *4*, 100127. https://doi.org/10.1016/j.envc.2021.100127

Olukanni, D. O., & Olatunji, T. O. (2018). Cassava waste management and biogas generation potential in selected local government areas in Ogun State, Nigeria. *Recycling*, *3*(4). https://doi.org/10.3390/recycling3040058

Owamah, H. I. (2014). Biosorptive removal of Pb(II) and Cu(II) from wastewater using activated carbon from cassava peels. *Journal of Material Cycles and Waste Management*, *16*(2), 347–358. https://doi.org/10.1007/s10163-013-0192-z

Poolprasert, P., & Chorchong, T. (2020). Preparation of activated carbon from cassava root for cadmium removal in aqueous solution sample. *Journal of Renewable Energy and Smart Grid Technology*, *15*(2), 26–38. https://ph01.tci-thaijo.org/index.php/RAST/article/view/243208

Rajeshwarisivaraj, Sivakumar, S., Senthilkumar, P., & Subburam, V. (2001). Carbon from Cassava peel, an agricultural waste, as an adsorbent in the removal of dyes and metal ions from aqueous solution. *Bioresource Technology*, *80*(3), 233–235. https://doi.org/10.1016/S0960-8524(00)00179-6

Schwantes, D., Gonçalves, A. C., Coelho, G. F., Campagnolo, M. A., Dragunski, D. C., Tarley, C. R. T., Miola, A. J., & Leismann, E. A. V. (2016). Chemical modifications of cassava peel as adsorbent material for metals ions from wastewater. *Journal of Chemistry*, *2016*. https://doi.org/10.1155/2016/3694174

Schwantes, D., Gonçalves, A. C., Fuentealba, D., Hornos Carneiro, M. F., Tarley, C. R. T., & Prete, M. C. (2022). Removal of chlorpyrifos from water using biosorbents derived from cassava peel, crambe meal, and pinus bark. *Chemical Engineering Research and Design*, *188*, 142–165. https://doi.org/10.1016/j.cherd.2022.09.033

Sharma, A., Mangla, D., Shehnaz, & Chaudhry, S. A. (2022, January). Recent advances in magnetic composites as adsorbents for wastewater remediation. *Journal of Environmental Management*, *306*, 114483. https://doi.org/10.1016/j.jenvman.2022.114483

Sithole, T., Nseke, J., Mashifana, T., Falayi, T., Dragoi, E. N., & Malenga, E. (2023). Neural network optimization during the purification of industrial effluents using steel slag: Kinetics and mechanism. *Environmental Technology and Innovation*, *30*, 103118. https://doi.org/10.1016/j.eti.2023.103118

Sowmya, P., Prakash, S., & Joseph, A. (2023, January). Adsorption of heavy metal ions by thiophene containing mesoporous polymers: An experimental and theoretical study. *Journal of Solid State Chemistry*, *320*, 123836. https://doi.org/10.1016/j.jssc.2023.123836

Ubalua, A. O. (2007). Cassava wastes: Treatment options and value addition alternatives. *African Journal of Biotechnology*, *6*(18), 2065–2073. https://doi.org/10.5897/ajb2007.000-2319

Wang, J., & Guo, X. (2020). Adsorption isotherm models: Classification, physical meaning, application and solving method. *Chemosphere*, *258*, 127279. https://doi.org/10.1016/j.chemosphere.2020.127279

Younis, M. Y., & Kayi, H. (2019). Artificial neural network prediction of sulfur content of diesel fuel from its physical properties. In *IOP conference series: Materials science and engineering* (Vol. 518, pp. 1–8). IOP Publishing. https://doi.org/10.1088/1757-899X/518/6/062008

Zhang, J. (2019, January). Physical insights into kinetic models of adsorption. *Separation and Purification Technology*, *229*, 115832. https://doi.org/10.1016/j.seppur.2019.115832

8 Future Perspectives on Waste Management

Freeman Ntuli

8.1 INTRODUCTION

Rapid urbanization and population growth, especially in sub-Saharan Africa (SSA), presents a lot of challenges in terms of environmental management and pollution control. It is estimated that about 50% of the global population growth will be in SSA, and the region's population is growing three times faster than the rest of the world.[1] Given the impact of climate change, declining economies, and infrastructure in the region, there is a need to adopt emerging advances in technology to help alleviate and manage the associated environmental challenges and resource utilization in a sustainable manner. The fourth industrial revolution has seen the emergence of advanced technologies driven by artificial intelligence (AI) as a possible tool to solve a number of industrial, societal, and economic challenges. Research has shown that AI can act as an enabler in 79% of all SDGs through harnessing technology to overcome most of the current limitations (Vinuesa et al., 2020). For SDG related to the environment, AI was found to be an enabler in 93% of the SDG targets through its ability to analyse large-scale interconnected databases to develop joint actions aimed at preserving the environment (Vinuesa et al., 2020).

8.2 CE IMPLEMENTATION IN SSA AND THE CHALLENGES

CE is premised on achieving three key objectives, namely:

- Elimination of waste and pollution through process design
- Keep materials and energy within the materials and energy loops
- Regeneration of natural systems

The principles of CE are not really new to SSA, and aspects of this have been practised in Africa, especially in the agricultural sector. The use of agricultural residues as livestock feed or as compost has been practised in most SSA countries to achieve the circularity of nutrients within agricultural operations. Repurposing and reuse of materials is also very common and is driven by economic constraints and the availability of cheap reused imported goods in the market. In most instances Africa has been viewed as a dumping ground for used goods such as vehicles, clothes, plastics, and electronics, thus extending the useful life of products that are obsolete in the developed world. As a consequence of this, the focus on the CE economy in Africa has been largely on the lower end of circularity, mainly waste management, rather

DOI: 10.1201/9781003377856-8

than designing out waste (Warner et al., 2021). A useful framework for understanding the CE strategies and targets was developed by Potting et al. (2017) and is illustrated in Figure 8.1.

Based on Figure 8.1, it can be seen that most initiatives for implementing the CE in SSA are mainly based on strategies R7, R6, R4, and R3, mainly focusing on extending the lifespan of products – that is, consumption. This is corroborated by a survey of 23 business case studies on the CE economy initiatives in Africa (Figure 8.2), which shows that most of the initiatives are focused on using waste as a resource (Warner et al., 2021). In order to achieve a full transition to CE, it has been demonstrated that policies and targets that focus on higher CE economy strategies (R0-Refuse, R1-Rethink, R2-Reduce) are more effective (Morseletto, 2020). Most of the current targets and policies in most parts of the world, including SSA, have mostly focused on waste management and resource conservation (R9-Recovery and R8-Recycle) combined with control of emissions to the environment (Morseletto, 2020). This approach has been found to not adequately promote a CE; thus, there is

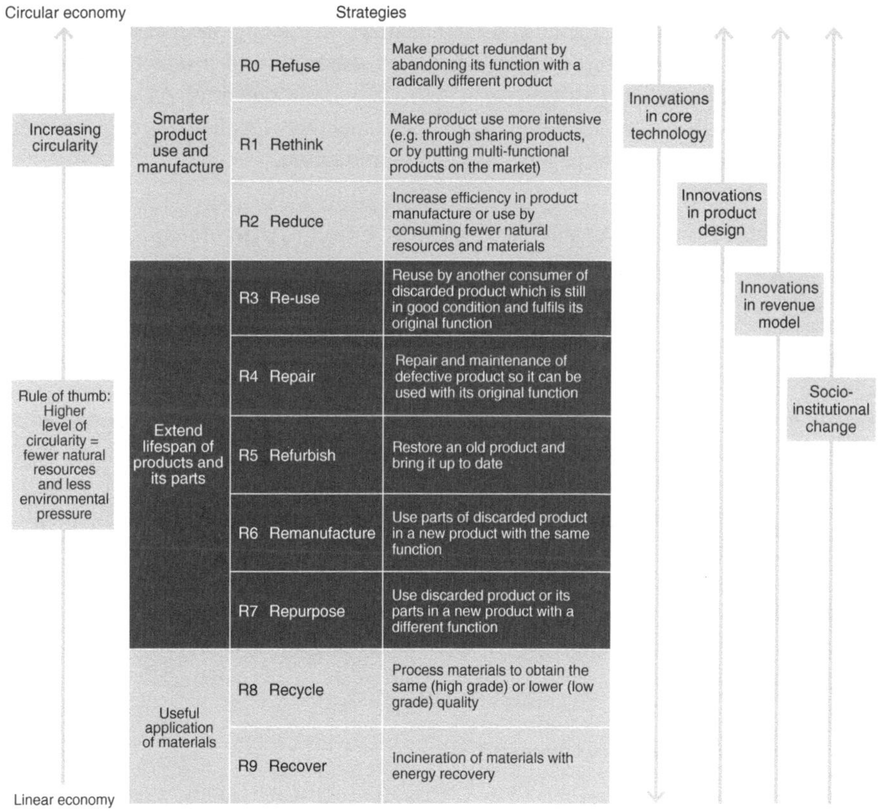

FIGURE 8.1 10 Rs of the CE.

Source: Adapted from (Potting et al., 2017).

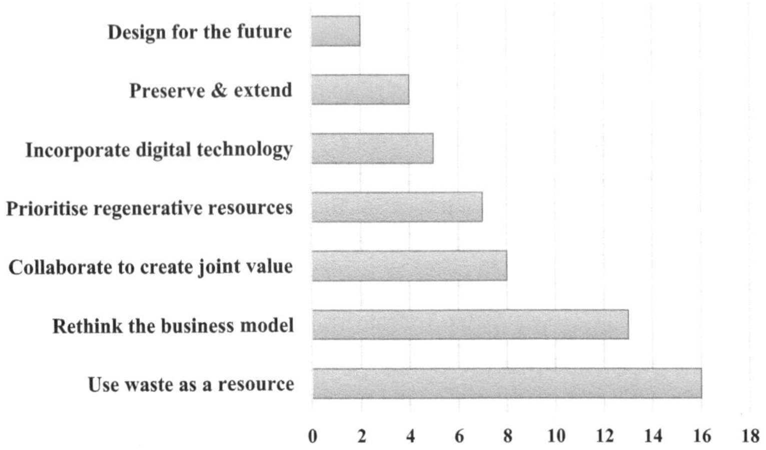

FIGURE 8.2 Business case studies for CE strategies in Africa.

a need to develop new strategies and targets aimed at closing production loops and maximizing retention of the economic value of materials and products in order to achieve a sustainable CE (Morseletto, 2020). The development of AI and machine learning tools will be key in achieving the CE objectives for the future, given their potential to achieve smarter product use and manufacture.

8.3 OVERVIEW OF AI TECHNOLOGIES IN ENVIRONMENTAL MANAGEMENT

AI is a branch of computer science that is concerned with the development of computer programmes that can mimic and go beyond human capabilities, such as problem-solving, identification, recognition, and understanding. Machine learning is a subcategory of AI concerned with the development of algorithms to automatically learn insights and recognize patterns from data, applying that learning to make increasingly better decisions. Deep learning is an advanced method of machine learning that uses large neural networks to analyse historical data to learn complex patterns and make predictions independent of human input.

Recently, there has been an increased interest and study into the use of AI and machine learning in addressing issues of environmental sustainability in a number of sectors, such as renewable energy, waste management, and recycling (Fan et al., 2023; Arashpour, 2023).

8.3.1 AI in Waste Management

Waste management using AI has been employed to address a number of problems in dealing with solid waste such as municipal waste management (Vyas et al., 2023) and marine plastics (Seyyedi et al., 2023). Currently, AI tools are mainly employed in solid waste management to predict the trends in the amount of waste generated, optimize refuse collection truck routes, navigate disposal facilities, and imitate waste transformation mechanisms (Vyas et al., 2023). The commonly modelled waste transformation processes are waste combustion and anaerobic digestion (McElhinney et al., 2022).

The most common AI tools for solid waste management used are linear regression (LR), genetic algorithm (GA), support vector machine (SVM), artificial neural networks (ANN), and decision trees (DT) (Vyas et al., 2023) with ANN being the most commonly used. ANN have been used in a number of applications ranging from predicting waste generation and bin levels, waste categorization, landfill leachate production, and material recovery. SVM has largely been used for bin-level prediction, waste generation rates, waste categorization, and waste recycling applications. LR is largely used to predict waste generation rates and landfill leachate production, however, it is not suitable for modelling non-linear data. DT have low computational complexity and is commonly used for predicting waste generation rates, waste classification, and waste generation behavioural patterns. GA has been widely used for the estimation of waste generation rates, waste categorization, and waste concentration as well as siting facility prognostication, evaluating heating of waste and biogas production.

Studies on the use of blockchain technology in waste management have also gained traction in recent years. A comprehensive review of the use of blockchain technology in waste management has been published by (Jiang et al., 2023). The integration of Internet of Things (IoT) and blockchain technology enables all waste management operations (i.e., waste generation, waste segregation, collection, and recycling) to be centrally managed, providing a basis for intelligent decision-making that promotes circular economy strategies (Jiang et al., 2023). Despite these advances in integrated IoT and blockchain technology, a number of challenges still need to be addressed, namely, data acquisition and transparency, development of human capital with a capability to understand and apply the technology, the cost versus the benefit of blockchain applications for sustainable waste management in remote or rural areas and systematic applications of blockchain technologies (Jiang et al., 2023).

8.3.2 AI in Wastewater Management

There are more AI applications in the field of wastewater treatment and management as compared to SWM. AI in wastewater treatment and management has been largely used for predicting water quality from wastewater treatment plants (WWTP) (Matheri et al., 2021), monitoring the performance of WWTP (Matheri et al., 2022), optimizing models and process parameters, controlling membrane fouling, and in anaerobic to predict gas production and to optimize the process (Singh et al., 2023; Dantas et al., 2023). The most commonly used AI tools in wastewater treatment and management are ANN, and the nine most commonly used ANNs in order of use are (Zhang et al., 2023):

1. Multilayer Perceptrons (MLP)
2. Radial Basis Function (RBF)
3. Long-/Short-Term Memories (LSTM)
4. Convolutional Neural Networks (CNNs)

5. Extreme Learning Machines (ELMs)
6. General Regression Neural Networks (GRNNs)
7. Recurrent Neural Networks (RNNs)
8. Artificial neural network hybridized with genetic algorithm (ANN-GA)
9. Artificial neural network hybridized with particle swarm optimization (ANN-PSO)

The most commonly used architecture is the feedforward neural network (FFNN) with a backpropagation training algorithm (Dantas et al., 2023). MLPs are a simple FFNN that has three layers and one hidden layer. If it has more than one hidden layer, it is referred to as a deep ANN. ELMs and RBFs are based on MLPs and have lower computational burden. RBFs use a radial basis activation function in the hidden layer. GRNNs are based on RBFs and are used to solve less complex problems. LSTM are mainly used to deal with time-series tasks since it is capable of learning sequences of events over a period of time and a version of RNNs. CNN are a recently applied feedforward ANN widely used in wastewater treatment. Hybrid neural networks that combine heuristic algorithms with ANN such as ANN-GA and ANN-PSO have better performance but have not been widely used in wastewater treatment (Zhang et al., 2023).

8.4 CONCLUSION

The use of emerging technologies such as AI has a great potential to address most of the challenges with achieving circular sustainable economies in SSA. However, there are still a number of drawbacks that still need to be addressed in the following areas:

- Development of legislation that fully addresses the requirements of a CE
- Setting up of legislative structure and their funding to ensure they have the capability to implement and monitor adherence to CE targets based on SDG
- Development of infrastructure to support the implementation of a digital economy
- Human capital development to support application of technology to achieve sustainable circular economies
- Regional integration within SSA to enable sharing of knowledge and technologies
- Legislation and legislative structures at the level of the African Union to promote SDG and sustainable circular economies

NOTE

1 https://foreignpolicy.com/2023/08/26/demographics-africa-sub-sahara-population-boom-growth-aging-gender-inequality-climate-change/

REFERENCES

Arashpour, M. (2023, May). AI explainability framework for environmental management research. *Journal of Environmental Management*, *342*, 118149. https://doi.org/10.1016/j.jenvman.2023.118149

Dantas, M. S., Christofaro, C., & Oliveira, S. C. (2023). Artificial neural networks for performance prediction of full-scale wastewater treatment plants: A systematic review. *Water Science and Technology*, *88*(6), 1447–1470. https://doi.org/10.2166/wst.2023.276

Fan, Z., Yan, Z., & Wen, S. (2023). Deep learning and artificial intelligence in sustainability: A review of SDGs, renewable energy, and environmental health. *Sustainability*, *15*(18), 13493. https://doi.org/10.3390/su151813493

Jiang, P., Zhang, L., You, S., Van Fan, Y., Tan, R. R., Klemeš, J. J., & You, F. (2023, June). Blockchain technology applications in waste management: Overview, challenges and opportunities. *Journal of Cleaner Production*, *421*. https://doi.org/10.1016/j.jclepro.2023.138466

Matheri, A. N., Mohamed, B., Ntuli, F., Nabadda, E., & Ngila, J. C. (2022). Sustainable circularity and intelligent data-driven operations and control of the wastewater treatment plant. *Physics and Chemistry of the Earth*, *126*. https://doi.org/10.1016/j.pce.2022.103152

Matheri, A. N., Ntuli, F., Ngila, J. C., Seodigeng, T., & Zvinowanda, C. (2021). Performance prediction of trace metals and cod in wastewater treatment using artificial neural network. *Computers and Chemical Engineering*, *149*. https://doi.org/10.1016/j.compchemeng.2021.107308

McElhinney, J. M. W. R., Catacutan, M. K., Mawart, A., Hasan, A., & Dias, J. (2022, April). Interfacing machine learning and microbial omics: A promising means to address environmental challenges. *Frontiers in Microbiology*, *13*. https://doi.org/10.3389/fmicb.2022.851450

Morseletto, P. (2020). Targets for a circular economy. *Resources, Conservation and Recycling*, *153*(October 2019), 104553. https://doi.org/10.1016/j.resconrec.2019.104553

Potting, J., Hekkert, M., Worrell, E., & Hanemaaijer, A. (2017). Circular economy: Measuring innovation in the product chain. *PBL Netherlands Environmental Assessment Agency*, *2544*, 42.

Seyyedi, S. R., Kowsari, E., Ramakrishna, S., Gheibi, M., & Chinnappan, A. (2023, July). Marine plastics, circular economy, and artificial intelligence: A comprehensive review of challenges, solutions, and policies. *Journal of Environmental Management*, *345*. https://doi.org/10.1016/j.jenvman.2023.118591

Singh, B. J., Chakraborty, A., & Sehgal, R. (2023, October). A systematic review of industrial wastewater management: Evaluating challenges and enablers. *Journal of Environmental Management*, *348*, 119230. https://doi.org/10.1016/j.jenvman.2023.119230

Vinuesa, R., Azizpour, H., Leite, I., Balaam, M., Dignum, V., Domisch, S., Felländer, A., Langhans, S. D., Tegmark, M., & Fuso Nerini, F. (2020). The role of artificial intelligence in achieving the sustainable development goals. *Nature Communications*, *11*(1), 1–10. https://doi.org/10.1038/s41467-019-14108-y

Vyas, S., Dhakar, K., Varjani, S., Singhania, R. R., Bhargava, P. C., Sindhu, R., Binod, P., Wong, J. W. C., & Bui, X. T. (2023, April). Solid waste management techniques powered by in-silico approaches with a special focus on municipal solid waste management: Research trends and challenges. *Science of the Total Environment*, *891*, 164344. https://doi.org/10.1016/j.scitotenv.2023.164344

Warner, H., Bingham, J., & Nartey, D. O. (2021). *The circular economy: Our journey in Africa so far contents* (pp. 1–48). https://irp-cdn.multiscreensite.com/40a0e554/files/uploaded/CEcasereport_Footprints.pdf

Zhang, S., Jin, Y., Chen, W., Wang, J., Wang, Y., & Ren, H. (2023, June). Artificial intelligence in wastewater treatment: A data-driven analysis of status and trends. *Chemosphere*, *336*, 139163. https://doi.org/10.1016/j.chemosphere.2023.139163

Index

Note: Page numbers in *italics* indicate figures, and page numbers in **bold** indicate tables in the text.

E

F

G

H

I

J

K

L

M

T

U

V

W

X

Z